PRAISE FOR

To Hate Like This Is to Be Happy Forever

"The book begs comparisons to Nick Hornby's *Fever Pitch* but owes its debt more than anything to maybe the greatest sports novel of all time, Frederick Exley's *Fan's Notes*. Blythe knows a thing or two about bitterness and longing, but tempers these twin emotions with intelligence and humor and even generosity. It goes without saying that if you hate Duke you should read this book. But if you love Duke, or you love basketball, or you just love the kind of sportswriting that comes along so rarely you can count the classics on one hand, you should read this book."

—Joel Lovell, *New York Times*

"Hilarious and remarkably wise . . . the sort of book you don't want to say too much about, for fear of spoiling the surprises."

—Charles Hirshberg, *Sports Illustrated*

"The best book on basketball I have ever read . . . destined to become a classic of sports literature. In North Carolina, the Duke-UNC rivalry is more important than nuclear disarmament, world poverty, or the fall of communism, and Will Blythe is its ideal chronicler. He writes about these border wars with high style, elegance, passion, and, of course, utter impartiality. I've been waiting my whole life to read this book."

—Pat Conroy, author of *The Prince of Tides* and *My Losing Season*

"A revelation. . . . An elegy to place and time and generation, it is also a story of fathers and sons and an elegant testament to the way pastimes are far more than ways to pass the time."

—Sara Nelson, *Publishers Weekly* (signature review)

"You don't have to be a Tar Heel or a Blue Devil to like *To Hate Like This* . . . because it's funny, perceptive, and smart."

—Jonathan Yardley, *Washington Post*

"Fans of college basketball will wish that all sportswriters possessed Blythe's ability to describe a game, to translate its tension, and to render its action. . . . He writes amusingly, self-deprecatingly, and often beautifully."

—Franklin Foer, *New York Times Book Review*

John Blythe

About the Author

WILL BLYTHE is the former literary editor of *Esquire*. A frequent contributor to the *New York Times Book Review*, he has written for the *New Yorker*, *Rolling Stone*, *Sports Illustrated*, *Elle*, and the *Oxford American*, and is the editor of the acclaimed book *Why I Write*. His work has been anthologized in *The Best American Short Stories* and *The Best American Sportswriting*. He grew up in Chapel Hill, North Carolina, and now lives in New York City.

To Hate Like This
Is to Be Happy Forever

To Hate Like This Is to Be Happy Forever

A Thoroughly Obsessive, Intermittently Uplifting, and Occasionally Unbiased Account of the Duke–North Carolina Basketball Rivalry

WILL BLYTHE

HARPER

NEW YORK · LONDON · TORONTO · SYDNEY

HARPER

Poems by James Applewhite and Mike Troy
are reproduced courtesy of the authors.

A hardcover edition of this book was published in 2006 by
HarperCollins Publishers.

HarperCollins books may be purchased for educational, business,
or sales promotional use. For information please write:
Special Markets Department, HarperCollins Publishers,
10 East 53rd Street, New York, NY 10022.

FIRST HARPER PAPERBACK PUBLISHED 2007.

Library of Congress Cataloging-in-Publication Data
is available upon request.

ISBN: 978-0-06-074023-8
ISBN-10: 0-06-074023-X

ISBN: 978-0-06-074024-5 (pbk.)
ISBN-10: 0-06-074024-8 (pbk.)

07   08   09   10   11   ❖/RRD   10   9   8   7   6   5   4   3   2   1

For my mother,
Gloria Nassif Blythe

And in memory of my father,
William Brevard Blythe

Contents

To Hate Like This
Is to Be Happy Forever

ONE

The Object of My Affliction

MY LITTLE DUKE PROBLEM

*"A man who lives, not by what he loves
but what he hates, is a sick man."*
—ARCHIBALD MACLEISH

I AM A SICK, SICK MAN. Not only am I consumed by hatred, I am delighted by it. I have done some checking into the matter and have discovered that the world's great religions and wisdom traditions tend to frown upon this.

Therefore, dear reader, I need your prayers. But even more than I do, the University of North Carolina's basketball team, the object of my obsession, needs them. Here is the depth of my sickness. It is several years back on a beautiful afternoon during basketball season. The cable is out. (Note to self: Kill Time Warner.) I am alone in my apartment in New York City, frantically hitting the refresh button on my computer screen, getting the updates of Carolina's shockingly bad performance against its archrival, Duke. So far, the Heels have shot 18 three-pointers and hit exactly five.

There is no end to my gloom. My father is in his grave, my marriage is kaput, my girlfriend is said to be in Miami (though what she is doing there I can't say, since we're not speaking), I have no income, and yet the thing that is driving me over the edge is a basketball game that I can't even see. North Carolina, my beloved North Carolina, is being brutalized by Duke, being outplayed by opponents who are too kind, too mannerly even to gloat. At least when your rival gloats, you know victory over you means something. Again and again, I hit the refresh button and am transported anew to a message board resounding with rending cries and moans from fellow Carolina obsessives, posting their dismay, miss by brutal miss. It's like tuning in to the distracted mutterings of old men alone on park benches, all over America. There are so many of us.

Grown men, presumably a lot like me, are spending their Sunday afternoon on the *Inside Carolina* message board, writing things like "I wanna hurl." BlueBlood cries, "My sixth-grade students are gonna rip me a new one."

While I myself never post, content to lurk, I've come to know the personalities of some of the posters. The clever but doomsaying Jeff Brown opened one season by writing an amusing, if despairing, list with the title "We Just Have a Few Minor Problems." A guy calling himself The Critic, who gets on my nerves with his constant pessimism, says, "Good night, folks."

I won't eat. I can't eat. Or maybe I should eat, since there is the possibility, faint perhaps, that through a small, apparently unconnected action, like ordering sushi from the Malaysian place down the street, I will change the karmic pattern at work in this game. It's chaos theory and not to be sniffed at. What's that classic example—a butterfly flaps its wings in the Amazon and two weeks later a major hurricane devastates the Bengal peninsula? Or, to put it in my terms, perhaps a tuna roll inside out will allow Jason Capel to actually hit a three-point shot. Maybe a bowl of *chirashi* will cause Brian Morrison to stop booting the ball out of bounds. And a nip of sake may teach goddamn Kris Lang (as he is known in my household) to hold on to the ball.

A former teacher of mine, a great scholar of Southern literature, believes that he can control games by maintaining the same posture throughout the contest and by doing some kind of weird voodoo gesture with his fingers every time an opposing player shoots a free throw. I'd rather try eating, so I order the sushi, but nothing works. Carolina is shooting 29 percent from the field, and Lang has exactly one rebound. Like a cancer patient, I continue to make bargains with God (who I am not sure even exists). But He must not be watching this game. Another Tar Heel three clangs off the rim. They lose by 26.

The message board erupts. Coolheel: "I could have shot 5 for 18 from 3 myself after having a six-pack, which was much needed to endure the flow of this stinker." UNCodeCorrect: "It's a huge shit sandwich and we're all going to have to take a bite."

Another fan writes, "I may have to sit out this year with a bad back," a pointed reminder of the hated Duke coach Mike Krzyzewski's condition during the 1994–95 season, when the Blue Devils suffered a beautifully horrible time of it, finishing 13 to 18. Overburdened, Krzyzewski took a leave of absence from coaching that year. Rumors swirled through the Research Triangle of the Duke coach in tears, huddled in his bedroom, wrapped in a bathrobe, muttering more inanities than Dick Vitale. Now, the normal human certainly would feel sympathy for a man in such pain. But I am a North Carolina fan and by definition, at least when it comes to Duke, not a normal man.

I came naturally by my prejudice in this matter from my father, William Brevard Blythe II. He was a lifelong North Carolinian, born in Mecklenburg County in 1928. His childhood during the Great Depression was paradisiacal, or so he portrayed it to his children, whom he liked to tease for being "city kids." (Until we got older and learned to hit back, we would actually cry when he called us this.) He had a pony and a dog; he roamed through the woods and the fields without supervision; he and a couple of friends had the initiative to build their own tennis court when they decided they wanted to learn the game. Like his father before him and like me after him, he graduated from the University of North Carolina. He could not understand why you might want to live in some other place. He loved his home state (trees, birds, soil, fish, crops, counties, ladies, barbecue) in a way that few people seem to love their home states anymore, home being a quaint, antique concept in a nomadic and upwardly mobile America.

My father used to love to tell a joke about Duke, or, more specifically, about the difference between the University of North Carolina, in our hometown of Chapel Hill, and Duke University, which was only about eight miles from our house but a universe away in our affections. In a sense, it was a riddle about the difference between being and seeming, and it went to the heart of my father's values. He would even tell the joke to international visitors to our home, who had no idea what he was talking about but usually chuckled valiantly at the punch line. I remember in particular one homesick, bespectacled Egyptian grad student whom we had signed up to host one semester and who sat at our table one Friday night eating country ham and biscuits, earnestly trying to understand our views on Duke and North Carolina. How well my father understood this poor man's homesickness, having once spent three months in Alexandria helping set up dialysis units, listening every day to the muezzins' calls to prayer ringing from the minarets, which seemed to be summoning him not to Mecca but back to North Carolina.

"How can you tell the difference between a Carolina man and a Duke man?" my father asked the Egyptian, who thought for a while and finally shrugged his shoulders in a gesture of defeat. This was not a fellow who had spent much time pondering the fiery temper of Art Heyman or the buttery jump shot of Walter Davis. And then, proudly answering himself, my father said, "A Duke man walks down the street like he owns the whole world. A Carolina man walks down the street like he doesn't give a damn."

The Egyptian student laughed conscientiously, looking from one member of my family to the other to see if he was doing the right thing. We nodded; yes, yes, you got it. "Oh, that is so funny," he said.

There are two kinds of Americans, it seems to me, with my father representing the first. Those for whom the word "home" summons up an actual place that is wood-smoke fragrant with memory and desire, a place that one has no choice but to proudly claim, even if it's a falling-down dogtrot shack, the place to which the compass always points, the place one visits in nightly dreams, the place to which one aims always to return, no matter how far off course the ship might drift.

And then there are those citizens for whom home is a more provisional notion—the house or apartment in which one sleeps at night, as if American life were an exhausting tour of duty, and home, no matter how splendid, equaled a mere rest stop on the Interstate of Personal Advancement. I am biased against this kind of nomadism, no matter how well upholstered the vehicles. The loss of adhesion to a particular place seems ruinous, and those without the first kind of home wander through our nation like the flesh eaters from *Night of the Living Dead*.

A great many of these flesh eaters pass through the pseudo-Gothic arches of Duke University, "pass through" being the relevant phrase. Duke is the university as launchpad, propelling its mostly out-of-state students into a stratosphere of success. While hardly opposed to individual achievement, North Carolina, by contrast, is the university as old home place, equally devoted to the values of community and local service. That, at least, is the mythology many of us swallowed as we grew up. So that when one roots for one team or another in the Duke–North Carolina rivalry, one is cheering as much for opposing concepts of American virtue as for adolescent geniuses of basketball.

The basketball rivalry between Duke and North Carolina has become the greatest rivalry in college athletics, and one of the greatest in all of sports. It is Ali versus Frazier, the Giants versus the Dodgers, the Red Sox versus the Yankees. Hell, it's bigger than that. This is the Democrats versus the Republicans, the Yankees versus the Confederates, Capitalism versus Communism. All right, okay, the Life Force versus the Death Instinct, Eros versus Thanatos. Is that big enough? This is a rivalry of such intensity, of such hatred, that otherwise reasonable adults attach to it all manner of political-philosophical baggage, some of which might even be true. I know because I'm one of them. During the 2004 presidential campaign, candidate John Edwards, the former senator from North Carolina, could not resist jumping into the fray when he told a reporter for *The Oregonian*, "I hate Duke basketball." Yes, cautious John Edwards, a man determined to wage a coast-to-coast campaign in which he alienates not a single voter. But there he was out in Oregon, watching television in the company of a reporter, and there was the Duke basketball team, trashing another overmatched opponent on national TV, and Candidate Edwards, a North Carolina law-school graduate, could not contain himself, could not choke back his distaste. A grown man who had otherwise put away childish things, he still had to say it, how he hates Duke basketball. Of course, he has his counterparts who feel similarly about North Carolina basketball. Why should this be so?

The answers have a lot to do with class and culture in the South, particularly in my native state, where both universities are located. Issues of identity—whether you see yourself as a populist or an elitist, as a local or an outsider, as public-minded or individually striving—get played out through allegiances to North Carolina's and Duke's basketball teams. And just as war, in Carl von Clausewitz's oft-quoted formulation, is a continuation of politics by other means, so basketball, in this case, is an act of war disguised as sport. The living and dying through one's allegiance to either Duke or Carolina is no less real for being enacted through play and fandom. One's psychic well-being hangs in the balance.

What is behind the hatred, the collective ferocity? The solution to that mystery begins not with basketball itself, but with the universities in both fact and perception. The schools stand a mere eight miles away from each other off 15-501, the heavily traveled thoroughfare between Chapel Hill and Durham. Put two different notions of the universe in the same atom, as it were, and there's bound to be disturbances at the molecular level. In quantum terms, it's matter meets antimatter. In basketball terms, it's Duke versus North Carolina. As Mike Krzyzewski once said, "Forget the Big Ten. . . . We share the same dry cleaners. . . . There is no other rivalry like this. It produces things, situations, feelings that you can't talk to other people about. Because they have no understanding of it." So while the two schools are geographically close, they're a world apart in just about every other way.

North Carolina is a public university, the oldest one in the country, chartered in 1789 and opened in 1795, when one presumably weary student by the name of Hinton James walked into town from New Hanover County on the coast in search of schooling. Duke is a private university, endowed in 1924 by the tobacco magnate James B. Duke, who gave his money in exchange for having a college previously known as Trinity named after him. He directed the school to erect Gothic-style buildings amid the pine forests and old tobacco fields, structures befitting a medieval university where scholars would go punting on willow-shaded rivers, not on a football field. By contrast, the North Carolina campus evolved in a more higgledy-piggledy fashion and features a heterodox assortment of architectural styles, ranging from a simple brick dormitory from the eighteenth century to hideous concrete-block fortresses from the 1970s. Throughout the years, an aesthetic of modesty has seemed to prevail. North Carolina draws a large share of its 15,961 students from within the state. Most of Duke's 6,347 students come from out of state, and many of them are accused by Carolina fans of being Ivy League wannabes who have fallen back on what their brethren call "the Harvard of the South."

In the Seventies, my snooty little liberal friends and I felt that we owned a devastating advantage in the Duke–Carolina argument by virtue of the fact that Richard Nixon had attended law school at Duke. We found this gem of a quotation from Tricky Dick that had a conclusive redolence, post-Watergate: "I always remember that whatever I have done in the past or may do in the future, Duke University is responsible in one way or another." This seemed to say it all, at least if you were a snooty young liberal.

Even in basketball, the universities find themselves on opposite sides of the political spectrum. A graduate of West Point, Krzyzewski is famously conservative, exalting business values, speaking to American companies about "winning in the corporate world." During the last U.S. Senate election in North Carolina, he got into a little hot water for hosting a fund-raiser on the Duke campus for the Republican candidate and eventual victor, Elizabeth Dole. By contrast, North Carolina's longtime coach, Dean Smith, the winningest coach in NCAA Division I history, has spoken out on behalf of liberal causes from the start of his career. He participated in the integration of Chapel Hill's public facilities in the early Sixties, supported a nuclear freeze in the early Eighties, opposes the death penalty, and was at times touted as a possible Democratic candidate for the U.S. Senate.

Both programs have been about as successful as college teams can be, North Carolina having won four national championships and Duke three. (North Carolina claims a fifth, from 1924, before the current playoff system began.) They are both among the top four teams in all-time victories, North Carolina having racked up 1,825, Duke 1,730. They began playing each other back in 1920, with North Carolina leading in the head-to-head victory count, 124 to 95. Each has employed some of the game's legendary coaches: Frank McGuire, Smith, and now Roy Williams at North Carolina; Vic Bubas and Krzyzewski at Duke. And both have enjoyed rosters resplendent with many of the game's greatest players: at North Carolina, Michael Jordan, Phil Ford, and James Worthy; at Duke, Art Heyman, Christian Laettner, and Grant Hill. No school in the country has signed more McDonald's All-Americans than Duke or Carolina—a measure of their allure to the nation's best high school players.

The minister who had presided over my father's funeral just moments earlier came up behind me and whispered in my ear. "UCLA by ten," he said. "But it's early." How he knew that I would want to hear a score update on that afternoon, my father's body barely lowered into the frozen winter ground, I don't know. I was standing in the receiving line in the church's fellowship hall with my mother and my brothers and my sister, greeting mourners, accepting their condolences, leaning forward to hear their memories of my father. His sudden death at 72 on December 21, 2000, was so surreal—he had seemed too ornery, too ribald, too full of plans to die—that all I wanted was to watch the Carolina–UCLA game. You can say it was my way of proclaiming, "Death be not proud." But my father would have called my bluff on that fine sentiment. "Acting that way about a basketball game at your age," he used to routinely upbraid me as I screamed and whined, "I thought you'd gotten over that."

"Me? What about Mama?" I would say, knowing this would irritate him even more.

"She's married to the television set," he'd say, exasperated.

"Please, let's not do this now," my mother would plead, not wishing to be distracted from the game on TV.

"You're all crazy," my father would say, settling down on the couch with the rest of us, defeated by the persistence of childhood passion in the grown men and women who mystifyingly happened to be his own wife and children. Gradually, however, he would get drawn into the game, and as his own anxieties about the outcome mounted, he would deflect them onto us. "Y'all stop that fussing," he'd respond when we yelled at referees and coaches. He often taunted us as we questioned a substitute or a defensive alignment, "Don't you think Dean Smith knows a little more about basketball than you and your mother?"

"Not about defending the three-pointer," we would yell. There were more than a few times when my mother and father ended up watching games on different televisions in different parts of the house.

At the church, I excused myself from the receiving line for a moment and found the minister, a kind and educated man who had been a great comfort to my family. Bob Dunham dealt with death as if it were an old and expected acquaintance who always showed up sooner or later. While my father lay in a coma, Bob sat in the hospital with us, talking about politics and eccentric church members and, yes, the weather. Bob prayed, too, but quietly, without show, as if prayer was a sort of desperate, inadequate communication that would have to do, given God's notorious silence.

"What is it now?" I asked.

"They've cut it to five," he said. Bob was clearly disappearing every now and then into the privacy of his office deep in the brick bowels of University Presbyterian Church, where evidently a television or a radio was broadcasting basketball even in the shadow of death.

"How much time?" I asked.

"Still the first half," he said.

"Who's playing well?"

"Forte's looking good," he said. If only all parsons ministered to their flocks like this, I might still be a churchgoer. I made it home that afternoon in time to catch most of the second half, which I watched huddled in a room by myself while the rest of the family and friends drank and ate in the living room.

Little more than a month later, I sat transfixed, watching the first Duke–Carolina game that I had ever seen without my father on this earth. North Carolina was ranked fourth in the country, Duke second. As you might expect, my father's feelings about Carolina basketball were complicated. If a speck of dirt, a smudge of red clay, a bird feather, or a weed came from North Carolina, he would love it. "Do you know where this weed is from?" he would have proudly said. "This weed is from North Carolina. Look at it. See how beautiful it is. They don't grow weeds like this in other states."

But he thought there was too much fuss made about sports in general and Carolina basketball in particular. "What will happen if they ever have a losing season?" he asked while considering the construction of the Dean E. Smith Center, better known as the Dean Dome, with its seating capacity of more than 21,000. Had he lived a season longer, he would have found out, and it wasn't pretty. Once he attended a game with my brother John and became irritated—you have to love this if you have the slightest bit of contrarian in you—because he felt the fans were making too much noise. According to my brother, he stood up and shouted, "Throw in the Christians," a reference to gladiators and coliseums and early Christian history that probably befuddled the few fans who actually heard the slender, bald-headed man in his blue blazer and khakis. "They're not the Christians," they might have answered. "They're the Demon Deacons."

And yet, my father hated Duke. Hated Duke with a passion that made him throw that blue blazer on the floor, that made him hoot and holler with the rest of us, only occasionally asking us to be a little more decorous. So that any bad behavior, any extravagant outbursts or strings of lowdown profanity we exhibited in cheering against Duke received a sort of papal indulgence from him.

Waiting for the game to begin, I had idly wondered with a child's mind whether my father might be able to affect the game from wherever he might be. Not that either he or I believed in life after death, at least not in that version presented by Christians and Muslims and various and sundry others: streets of gold, 72 virgins, white robes, fleecy clouds. Give me an October field in late afternoon, a drive between two beloved places, a hand on my knee, a plentiful solitude. We both wanted to believe that life went on in some fashion and that we might see our kin again. (There are a few people I'm sure we had no desire to reconnect with.) After my grandfather—his father—died, my father said by way of comment on Pappy's simple faith in the next world, "I'm a scientist." He meant he couldn't see how the facts as he knew them would allow for his fondest hopes. That said, I wondered still if he were comfortably ensconced in some other universe, a mere wormhole away, and if he might not be able to confound Jason Williams's jump shot, or at least stop the referee from giving Shane Battier license to flop (he was widely known as Floppier by Carolina fans) and draw a charge from an opponent.

Certainly we could have used such telekinetic intervention. The game was being played at Duke's Cameron Indoor Stadium, a hellish hive of vituperation posing as wit. I was alternately envious of Duke's Nuremberg-rally cheers and appalled at the smug mockery that nincompoops like Dick Vitale claimed as evidence of high SATs and a predictor of future success. The Carolina players were introduced.

"At center, at seven feet, from Greensboro, North Carolina, Brendan Haywood," the announcer announced.

"Hi, Brendan, you suck!" the Duke students chanted en masse.

The Tar Heels' superb if volatile shooting guard Joe Forte was introduced. The students greeted him in unison. "Hi, Joe, you suck!"

And so it went on down the starting five. The only thing was, Joe didn't suck that night. He didn't come close to sucking. If Joe sucked, then sucking should be required. Sucking should be part of the curriculum. Sucking should be a station at basketball camp. Because Joe was sucking as in sucking up every rebound in sight, driving and shooting and rebounding as if he were at home, as if the game were being held in the gym at DeMatha Catholic High School in Hyattsville, Maryland, where Joe had honed his throwback skills of midrange shooting—16-foot jumpers falling into the net like groceries plopped into a bag by a pimply checkout boy. He wasn't a great leaper. His game had a 1950s Midwestern element to it, a fundamental rightness.

Normally, the Duke players displayed good fundamentals themselves. Too often, actually, they reminded me of past Carolina teams—getting fouled more than their opponent and going to the free-throw line, where they nailed shot after shot, making more than the opposition even attempted. Nerves of Teflon. But tonight, to my astonishment (could it have been my father messing around in his alternate dimension? Did they have ESPN there?), the Duke players were shanking free throws. Shanking them, clanging them, spinning them off the rim like a rec-league team of eight-year-olds. Only Jason Williams was keeping Duke in the game, hitting outside shots and driving layups.

At halftime the score was 41 to 34 in favor of Carolina. At first I had only wanted the Tar Heels to make it a good game, keep it close, play hard, not embarrass themselves, given that they had lost five straight in the rivalry. Then I wanted them to make it a very, very good game. But screw that: Now I wanted what I always wanted and had been too worried, too fearful to ask of the basketball gods: total, unconditional, World War II–style victory. A loss would be devastating after having played so well, after having gone into Cameron and executed with such control and precision. Even the relatively lead-footed Jason Capel, Carolina's six-eight forward, had landed a shrieking, gravity-defying dunk on the befuddled heads of the Blue Devils. For a moment I wondered which player had soared like that. Capel? I respected his ardor in going after rebounds, but I was more used to seeing him pump-fake three times in a row and then get the ball blocked off the top of his modestly Afroed head. I imagined that he had to take a pick at halftimes and fluff the ball imprint out of his hair.

But tonight he seemed transported to another quantum level. Three-pointers, dunks, fist shaking, screaming. Capel liked to scream. Often, he would scream and pound his chest and make these weird horns over his head with his arms even when Carolina was losing by double digits. Fortunately, that wasn't the case tonight. This was good to see.

So why, as the second half unfolded, was I feeling dread? The first answer is simple: I always feel dread when Carolina is playing Duke. Normally, the Blue Devils are a good team, and for the last few years, they've too often been a great team. So there is always the possibility of losing and the psychic turmoil that roils my world in the wake of that. But even if Duke has a bad team—as they did in 1995, when the Tar Heels and the Blue Devils played one of the greatest games in the series—the dread is there, because there is nothing worse than losing to an inferior Duke team. So there is no way I can win. Except by Carolina's winning. And not just winning occasionally, but by winning every game.

The second reason I was hearing the sneaker squeak of doom was that Duke was too good a team this season not to make a comeback. With 1:28 left in the game, Carolina was leading by a mere five points, 76 to 71. The lionized Battier fouled Tar Heels point guard Ronald Curry near half-court, and Curry went down hard, holding his left thigh. I hate to see players injured, especially North Carolina players, but if anyone was going to be injured before having to shoot two free throws, I was glad that it was Ronald Curry, whose charity-stripe percentage stood at a resoundingly awful 41 percent. Now, whether Curry was really injured— can a thigh bruise stop you from shooting free throws?—I'll leave to the Duke conspiracy theorists. Clearly, Krzyzewski thought that Curry's injury was about as credible as the Warren Commission Report. He leaped off the bench and yelled at the officials, though to no avail. Carolina coach Doherty substituted Max Owens (who just happened to be an 80 percent shooter from the line) for Curry. Owens sank the free throws, and Carolina now enjoyed a seven-point lead.

It wouldn't last. Down by three with perhaps five seconds left, Duke's Mike Dunleavy, a gangly Ichabod Crane of a player, took an awkward, fading three-pointer. I had been screaming, "No threes! No threes!" for all the Upper West Side of Manhattan—perhaps the entire universe (was my father listening?)—to hear. The shot arched into the air with dispassion. So calmly did the ball float through the air. So lovely was its migration. As an aesthetic phenomenon, it really couldn't have been topped, parabolas always possessing such spellbinding curvature. And damned if the shot didn't fall cleanly through the hoop and stab me in the heart like the sneakiest cheatin' girlfriend.

The crowd exploded. How I wish it really had exploded—you know, blue-painted body parts shooting through the air, cheerleaders spiraling above the city of Durham, all those obnoxious students and that out-of-state arrogance disappearing in one bright blast. No more references by Dook, er, Dick Vitale to the future doctors and lawyers acting like animals before they headed off to Harvard Medical School and Wall Street, the blue paint still slathered on their bodies as if they were lost tribes of the Amazon. Forgive me, I know this is wrong. But, still . . .

With 3.9 seconds left, Doherty called time-out and gathered his players around him. They seemed remarkably relaxed. That may have been because earlier, in an effort to keep his players loose, Doherty had told them, "Duke has the ugliest cheerleaders in the ACC." Later, this comment leaked out to the press, and Doherty was forced to apologize. I don't recall his exact words of regret, but the remark certainly worked for his players, who, mysteriously to the observant fan, seemed to be laughing in the huddle.

Carolina had designed a play for—who else?—Forte, who by this point had collected 24 points and 16 rebounds, an astounding figure for a guard. The inbounds pass was supposed to go to him as he ran down the sideline like a wide receiver on a fly pattern. He was supposed to catch it and get the best shot he could. A reasonable if haphazard plan, given that Duke could be expected to key its defense on Forte, which was exactly what it did. Thus, when the harassed guard got the ball, he had no choice but to fling it downcourt to, of all players, Brendan Haywood.

At the same moment that the ball reached Haywood, so did Battier, dubbed "the Golden Boy" by the wry Lang, who, in contrast to Battier, possessed a lot of brass but not much touch. Whistle. Foul. Amazingly called on the Golden Boy himself by referee Mike Wood. Duke fans must have felt that Wood had some nerve. Twenty-two feet from the goal, 1.2 seconds left in the game. "Just say I was a defensive back and got called for pass interference," the ever-glib Battier would say later. "It was just basketball. I was trying to bat the ball away to get to overtime. I wasn't trying for a steal. I collided with Brendan, and I guess I lost, because he got the call."

Haywood was an extremely good shooter when he was within about two feet of the basket, which is to say he was very good at dunking the ball, screaming, hanging on the rim, and not much else. At the free-throw line he was an ironworker, averaging only 48 percent. When he was a freshman, he missed a pair of free throws late, and Duke snaked past the Tar Heels by two points. "I was definitely remembering those free throws," Haywood confessed. "That was the first thing that was running through my mind."

Agony. Agony. Agony.

Haywood stepped to the line and fired. Swish. He calmly sighted one more time and shot. Swish. "I just had to focus and go ahead and follow through and think of my mechanics," he said to reporters later. "Luckily, they went in for me. There's a lot to be said about luck." His gratitude to luck—the unseen forces at work in the universe—struck me as just and gratifying and the sort of sentiment rarely expressed by athletes, who, like Republicans, tend to see success not as remarkably provisional but as the confirmation of their rightful place in the world. "I didn't want to see myself on TV as part of an instant classic," Haywood added.

Duke had one last chance. Chris Duhon hoisted a shot from near midcourt, and as time expired, the ball bounded off the back rim. Too close for comfort, but off just enough for jubilation. North Carolina triumphed, 85 to 83. I'm not the type of guy to point up to the sky at a dead relative the way many athletes do these days. But I wondered. Or, anyway, I wanted to wonder.

From time to time I have felt silly about this devotion to a college team and the concomitant hatred of its rival. Here I am, a grown man, huddled in front of a TV, hiding out from the world from November to April, watching students battle each other in games that shouldn't mean more to me than to them. Right?

Not long ago, as I watched Carolina endure a particularly ugly sequence against Duke, I scared my girlfriend's nine-year-old son, Harry. (I had already terrified the dog, the beloved Gracie, who had fled into the bathroom to avoid my raving.) Duke's Dahntay Jones had just driven home a particularly obnoxious dunk and was now flexing his muscles like an insane bodybuilder. Was there no justice in the universe? Where was God?

I pounded my hand on the coffee table, stomped my feet on the floor, and exclaimed, with extreme eloquence, "Shit, hell, piss, damn it! And don't say what I just said, Harry!" Indeed, I felt proud of myself that I had limited my profanity to just these few words. A virtual Zen master of self-control.

Harry, who had been watching me watch the game, asked, "Why do you have to get so mad?" Normally, he would have delighted in an adult's swearing. But now he was edging backward across the room, the way people will when you have a gun pointed at them. His eyes were wide.

"Because I hate Duke," I explained.

"Why do you hate them?" he asked.

Here I hesitated. A young boy had asked me a guileless question, and he needed an adult response. "Well, that's an interesting question," I told him, channeling Mister Rogers, "and it deserves an honest answer." I paused for a moment, as I had seen his mother do when addressing an earnest inquiry by her son. Children are our future. We must teach them well, even when it is hard.

"The truth is they are terrible people," I told him. "Detestable."

"All of them?" he asked.

"Every last one of them," I said. "Especially the coach."

"I hate them, too," Harry said, settling in next to me on the couch. And thus was born another soldier in the war. On the door of his room hung a chalkboard for self-expression, and I was pleased to note that now, scrawled in his child's hand (with no assistance or prodding from me) was the unimpeachable sentiment, *NO DUKE FANS ALLOWED IN HERE.*

THE FAN IN THE SPORTS BAR

Let's analyze the author on a November day. On one such afternoon, his beloved Tar Heels win. On the same day, everything is otherwise identical, except that his team loses. What is the difference between winning and losing as it relates to the partisan soul?

On the afternoon North Carolina wins, he is kinder. Before he leaves the sports bar where he watched the game, he tells the bartender a dumb joke about a cow and a chicken, and the bartender laughs, not because the joke is funny but because his customer is spilling over with joviality and goodwill. He slaps hands with the black bouncer at the door. How the races love each other! He calls his girlfriend to simply tell her he was thinking of her. "Did they win?" she asks. "They won," he says. "Thank goodness," she says.

He sees the pallid New York light filling the rifts between buildings in more ebullient metaphorical terms—it is the wilderness light of the West, pouring into remote canyons, anointing the heads of explorers and lovers bouncing down rivers in rubber rafts. He smiles at a mother wheeling her twins in one of those 18-wheeler strollers that he so hates for their presumption of public space. He transmits sweetness and light in all directions. *Hello, you lovely East Side shrew! Let me pin myself against this building so that you may pass with that truckload of future Nobelists. Good afternoon, officer! Have you lost weight? Hi, kids! Would you mind turning up the volume on that new Kelly Clarkson single?* It's as if he had gone to church, made his confession, and come out on fire with happiness, an apostle of possibility. His church being basketball, he sort of has. He is renewed. He tips the taxi driver three dollars on a three-dollar fare. Victory has sweetened him into generosity and expanse.

Had the Tar Heels lost, oh, it would have been a different story. He would have drawn inward, pondering the doom-laden trajectory of his life. He would have sat alone at the bar trying to figure out what he was doing sitting alone at the bar. He would have hated the couple entwined around themselves in the corner, the girl with her stupid Brooklyn Dodgers cap, the guy with his designer tattoos. They're not even watching the game! Even when he was once similarly entwined—and how long ago that was—he stole glances at the game. Or suggested to his girl that they resume their necking at halftime. Maybe *that* was why he was now sitting alone at the bar.

He would have debated the size of the tip to leave the bartender, a ferocious internal squabble in which he would calculate how many more newspapers he could buy over the years if he left only a buck. And anyway, wouldn't the bartender interpret a large tip as evidence of deep loneliness that he was trying to buy his way out of? And even if he had left a decent tip, the reckoning of such with a miser's precision would have left him so spiritually depleted that maybe it would be better to leave nothing at all. The bartender would not have even noticed him leaving. The bouncer would have eyed him like a robber. He would have skulked out onto the street, invisible as a ghost, and seen the thin light petering out by the time it reached him. It would be all gone by four at the latest. Winter in New York sucked the spirit from a loser like an olive being pitted. The light is too soon leached out of the day and the soul. A foreshadowing of the descent of that eternal night. It was coming, closing in. Soon, soon the shot clock would run out. And he, a middle-aged man—middle-aged!—is ruminating exactly like a middle-aged man.

And that would be the supreme insult: His situation (middle-aged, steeped in the bitter tea of loss right on schedule) was a cliché! It lacked originality. Talk about pouring salt on a wound! He would head home to a clichéd bachelor's mess, the bed unmade, covered with books and spare change, the dust balls rolling down the hallway like tumbleweeds in a ghost town, the top of the couch shredded by his former wife's cat, the cat herself now gone from this earth, having never accumulated a name other than "the cat." And something like the cat's fate seemed in store for him: death arriving with no distinction for a fellow who would be known (if known at all) to coming generations simply as "the man." No child, no grandchild following after him to mourn. No one to receive the voluminous record of his life that he had saved as if there were going to be heirs, and now stored in the back of the closet in cardboard boxes begged from liquor stores. To whom does an isolate bequeath old love letters? Who will see that his grave's swept clean?

And yet, if only some player on the North Carolina basketball team, let's say Adam Boone back in that dreadful 8-and-20 day, if only Adam Boone, that future doctor, could have dribbled the ball into the front court without being trapped and panicking, this poor loner watching Adam Boone would have risen from despair to joy.

The author had arrived in the middle of his life with his happiness dependent on the fortunes of the home team in a place he had not lived for many years. How did this happen?

I lived in New York, quite happily, but the spiritual homeland of the Blythes was North Carolina. We were raised to love our state. Do people still love their states like that? Do you Dakotans, North or South, feel that there may be places more celebrated but none as beautiful or legendary as your own? That field behind your house? That snowbound city across the tracks? That prairie where you stood with your father and watched a comet overhead? Are they not the finest places on earth? In my 25 years away from North Carolina, I had watched my home from afar like a man following an old love down the years, typing her name into Google, piecing together the clues of her life without him, that posthumous existence.

In the fall of 2004, I had neither job nor wife nor children, and with the cat now departed for the Elysian Fields, where I hoped there were a lot of fat pigeons and slow mice, I was at liberty to make a job out of what I would otherwise be doing for free: watching and talking about basketball. I had no career, but that was all right by me. I didn't want one. Careers are drastically overrated, and since I no longer had one, it was probably better not to want one. I decided I would go back home, to the roots of my obsession.

So I would go south to my native state of North Carolina, into the heart of the Duke–North Carolina rivalry, its original territory. I would chart the fabled realms of a collective obsession. I would visit the principals and the mythic figures—the coaches, the players, the reporters, the fans, and my mother. (Oh, yes, my mother was a principal, and mythic, as mothers tend to be.)

As part of my journey, I would even cross over the invisible border that separated Duke University from the rest of North Carolina—go behind enemy lines, if you will—and conduct an extensive surveillance mission to see how the other half lived. And to record what would happen to the hatred that I nursed within me, a most ravenous and opinionated of beasts.

THE CONSOLATIONS OF UMA THURMAN

But before I headed south, I had to finish a couple of jobs in New York, one of which was a magazine story about Uma Thurman. We were supposed to meet on the evening of November 22, the first night of the Maui Invitational, in which the Tar Heels were playing Brigham Young.

The game was scheduled to be televised. I had to ask Thurman's people to move our interview up due to unforeseen circumstances.

Instead, we met for breakfast on the morning after the Santa Clara game, the first contest of the season, which I was unable to get on TV. I had the game figured for a rout, anyway. Carolina was loaded—all five starters returning from the previous season, the first under Roy Williams, in which the team had finished 19 and 11, making it to the second round of the NCAAs. The Tar Heels had also added an extraordinary freshman from Bremerton, Washington, the six-nine Marvin Williams.

I was scanning the box score to the Santa Clara game when Thurman arrived. She was lovely, a long pink scarf draped around her swan's neck.

"This is unbelievable," I said.

Out on the West Coast, North Carolina had lost to Santa Clara, 77 to 66. I explained the shocking nature of the upset to Thurman, the way it had ripped a hole in my sense of normalcy, toyed with my expectations, screwed my sense that the world would deliver justice and satisfaction. It should have been a blowout, I said, and it was—for Santa Clara. The point guard Raymond Felton didn't play because of a one-game suspension for having competed in an unsanctioned summer game. Rashad McCants had scored 20, hitting on four of six threes, Sean May snared 19 points and nine rebounds, but nobody else did much of anything. Melvin Scott had gone one of five on three-pointers. Some Santa Clara guy had racked up 26 points, and Sean May had blamed this on the fact that the fellow was left-handed! *Sports Illustrated* had ranked the Tar Heels number one in the preseason, but that wasn't going to last. Some fans were blaming the vaunted *SI* curse. To me, the performance felt like many over the last few years—sloppy play, bad defense, an odd loss to an odd team. Roy Williams was mad, and so was I.

Thurman was very kind and tried to keep my spirits up. She was going through some hard times herself, like the tag end of her marriage to Ethan Hawke, which I had to admit was painful, having myself gone through the dissolution of a long matrimony. On the other hand, that was just life. Theoretically, you could always get another spouse. This was basketball. You couldn't get another season with Raymond Felton,

Sean May, Marvin Williams, and Rashad McCants on the same team. No way.

"I'm so sorry," Thurman said.

"Let's not talk about it," I said.

"It might make you feel better," she said.

"You don't believe that," I said.

"Not really."

"How's *your* love life?"

"Let's not talk about it."

She ordered eggs Benedict, home fries included, and ate like a horse, which I admired. We were both suffering, but we tried to make it through the best we could.

BALL

UNC versus Indiana

December 1, 2004

For some reason, I had chosen to watch tonight's game at my girlfriend's house, where her son, Harry, was now banging a soccer ball against the back of my head in a steady, not unpleasant rhythm. He had recently discovered how much fun it was to throw, kick, and dodge a ball. And that what was fun in Washington Square Park was even more enjoyable in an apartment, where there was always the possibility of an exciting, unexpected, and altogether illegal collision between ball and household object. His mother had laid down the law about throwing the ball inside, but she often violated her own commandment by playing with Harry a suspenseful game of chuck-the-ball-across-the-bed-but-watch-the-overhead-lamp-and-the-lamp-on-the-bedside-table. These were apparently the sorts of things a modern-day mother and her son did. Many items in the house had met their demise in this season of the ball—lamps, bowls, wineglasses, and now, it seemed, my head. The last was probably no great loss. In a stupor of absentminded pleasure, Harry kept hitting me with the ball as I sat in a stupor of my own, studying the North Carolina–Indiana game on the tube.

"Is there a four-point shot?" he asked.

"N-n-n-o-o-o," I said, my voice trembling the way it will when a ball is vibrating your head at minute and regular intervals.

"If you don't want him to hit your head with the ball," my girlfriend said, "just tell him."

"Y-e-e-es-s-s-s, I-I-I w-w-w-w-l-l-l-l-l-l-l," I said.

"He doesn't mind," Harry said.

Ordinarily, this was true. But at the moment, things weren't going well. The Hoosiers had just scored on a backdoor cut, and I was beginning to feel like the team needed my full attention. Rashad McCants and Raymond Felton were playing well for the Tar Heels, but Sean May was pressing too hard in his return to his hometown of Bloomington, Indiana.

Carolina was having a hard time running in this game and could never pull away. In the second half, Marshall Strickland landed a three to bring Indiana to within one, at 34 to 33.

"S-s-s-t-t-o-o-p-p n-n-o-o-w-w!" I snapped at Harry. Despite the wah-wahing of my voice, he could tell I was serious. He went in search of his mother, hoping for some over-the-bed tossing. I had to do something to change the mojo. With a slender lead like this, a spectator really had to focus. On the bench, Roy Williams sat stone-still, posture rigid, as if he were using my tried-and-true method of sitting absolutely still in order to maintain a precious lead. I flew planes through turbulence in identical fashion, having discovered that as long as I kept the same stern posture, the plane stayed aloft.

And lo, Williams's body voodoo seemed to be working. Or was it mine? Didn't matter. McCants made a steal, Indiana fouled him hard, and he hit two free throws. UNC crept out to a three-point lead. The Hoosiers' momentum appeared thwarted by the play. McCants gave May a long stare, as if to say, *We'll get this for you. Don't worry.* Then on the next offensive series, he drove for a layup. Then Felton piled on with a three! Then McCants stroked a three-ball home. Indiana called time-out with the Heels sporting an 11-point lead.

When play resumed, May finally broke through, dunking the ball and being fouled at the same time. He exhaled visibly. He had scored a grand total of three points, but his timing was impeccable. Unfortunately, he got whistled for his fourth foul on an absurd call shortly thereafter. Exiled to the bench, he stood, whipping a towel overhead, rooting his teammates onward. He wanted this one bad. And for all practical purposes, it looked like he would get it.

But maybe not. Indiana rallied. Then May returned to the game. And magically, as if in cosmic retribution against the hometown haters, the ball squirted out of a scrum directly into the North Carolina center's hands and he scored. And once again was fouled. Carolina claimed a seven-point lead at 53 to 46 with just over four minutes left.

Unfortunately for the domestic tranquility of our tiny country, Indiana rallied yet again. "Can you guys keep it down in there?" I yelled at Harry and his mother. A concerned silence answered back. UNC had already committed 21 turnovers, with Felton having incurred seven by himself. The Tar Heels were slinging the ball all over the court. Roy was mad now, and when he got mad he stomped around like a Pentecostal preacher, only the tongue he was speaking did not appear to originate with the Holy Spirit.

As the time dwindled, Indiana kept knocking down threes. It was amazing, it was absurd, it was making me sick to death with worry. The commentators had already named Rashad McCants the PlayStation Player of the Game, and I wished they hadn't. For such hubris, we were paying a heavy price. With 46 seconds left, Indiana had trimmed the once imposing Tar Heel lead to six.

Sitting stock-still was no longer working. At moments of extremity, it always seemed wise to tinker with karma. Move things around a bit, change the narrative. "Harry," I called. "Get in here. And bring that ball."

"Do you want me to hit you in the head again?" Harry asked. He was a very intuitive boy.

"Yes," I said. "Fire away."

This immediately paid dividends. Sean May put in two free throws with 29 seconds left while the Indiana fans chanted, "Traitor, traitor." These points should have been enough to distance the Tar Heels from potential catastrophe as long as Indiana didn't hit another three. But damned if they didn't. And even though Felton responded with two free throws of his own, the Hoosiers hit yet *another* three to cut the Carolina lead to five, at

68 to 63. Bump, bump, bump went the ball on my skull. Although less than 20 seconds were left, anything could happen. Bump, bump, bump.

But what did happen was that Melvin Scott received the ball and was fouled and with chilly aplomb finished the game off with two free throws. Final: North Carolina, 70; Indiana, 63.

Harry fired another ball off the back of my head. I wanted to laugh and cry at the same time. "Do it again," I told him. And he did.

THE DREAM OF MIKE KRZYZEWSKI

A few nights before I went south, I dreamed about Mike Krzyzewski. How shall I put this? He and I fell in love. We stared into each other's eyes. Coach K understood me, I understood him. We lounged in the middle of the Duke campus on a glorious spring afternoon, flowers blooming among those faux-Gothic archways, a swirl of students floating past. It was all platonic, mind you, but nonetheless sweet with brotherly love. It could have been a lost episode of *Brideshead Revisited*.

When I awoke to the arid silence of four o'clock in the morning, I felt ashamed. My mind whirled, trapped inside itself like a CD racing around a spindle. To whom could I confess? What did the dream mean? And what did it mean that the figures frolicking in the green campus of my unconscious emanated from a sports rivalry?

I arranged an emergency meeting with a psychiatrist, and in the non-judgmental haven of her Manhattan office, I recounted the Krzyzewski dream. The psychiatrist listened intently, her head tilted slightly forward.

"What do *you* think it means?" she asked me, the standard catch-and-release technique of fly fishermen and psychiatrists.

I thought hard. Or, I feigned a look of thinking hard. Had there been a clock audible in the room, we would have heard it ticking.

"Hell if I know," I finally said. "But it sure worries me."

"It doesn't mean you're gay," she said. "If that's what you were worrying about."

"That never crossed my mind," I said. "Jeez."

"It's your shadow side," she said. "The side of you you're unwilling to acknowledge that is part of you. It thrills you. It's aggressive and competitive and ferocious and erotic."

"But I know I'm aggressive," I protested. "So does everybody else. And erotic? Wow. It's Duke, Doctor! Mike Krzyzewski! This would be like a Soviet dissident dreaming that he fell in love with Zbigniew Brzezinski."

She could tell I was upset, and, being a kind person, one of the kindest I knew in New York (she was a Christian, too, who believed in God and life after death—how had I ended up in the office of what surely must have been the only Freudian in New York who was also a believer?), she soothed me with words unusual for a shrink. "It's only a dream," she said. "But I want you to think some more on it. And we'll discuss it the next time you come in."

"That won't be for a while," I said. "I'm headed south." What *did* it mean? I shivered as I went out into the December streets of Manhattan, but not from the cold.

TWO

Heading South

SOUTH

SO SOUTH I WENT, south toward home, my mood rising as my latitude sank and I crossed those invisible lines, riffling down the Eastern Seaboard. *Whump, whump, whump.* Somewhere around Elkton, Maryland, I sliced through the Mason-Dixon Line, that old demarcation of North and South. Then Interstate 95 curved like the bend in a river through Baltimore, and it was here, in the so-called northernmost city of the South and the southernmost city of the North, that I began to feel I'd reached familiar ground. Here people's manners tended to be the ones with which I'd grown up. Almost imperceptibly, my Southern accent began to thicken.

Identity struck me as promiscuous in our country, a function of nomadism and speed, our love of upward mobility and new beginnings, a romance enshrined in the lottery. It seemed that one might choose whom one was to be. And yet, as I headed farther south, an involuntary persona, rooted deeper than choice or will, began to take over. Even on Interstate 95, buffeted by speeding trucks and SUVs, my relation with the world was turning from headlong to sideways. This must be why the South once featured so many porches, I thought. Because they provided the best opportunity for being sideways. And the Blythe family had long been masters of sitting on the porch, able to talk for hours quite happily about nothing. If we had a family philosophy, which I don't think we did, it would have advocated the virtue of sitting.

Which was a good thing, as I was soon stuck in traffic on the nightmare radius of Washington, D.C., trapped in the daily escape of bureaucrats and lawyers and office workers from that lovely marble city to the suburbs of northern Virginia. Ahead of me was a Quebecois family with three beautiful girls who wrote messages in idiot's French on pads and pressed them to the rear window for me to read. *Bonjour! Ça va? Merci!* I waved and shrugged in what I thought was a Gallic manner, causing them to laugh, perhaps derisively. I didn't care. Around Fredericksburg, I finally lost contact with my lovely tutors and continued south.

I picked up I-85 at Petersburg, Virginia, within sight of the old brick buildings of that city, the tin roofs and the polite, rectilinear angles that were the architectural legacy of a more formal time. Christmas candles flickered in the windows. I loved the stretch of road between here and South Hill, Virginia. You could really pound the accelerator down on that tree-shrouded corridor. On the radio, I began to pick up a winter's-night collage of basketball games. Around the North Carolina border, I caught the familiar voice of Woody Durham, the longtime broadcaster of UNC games, growing louder like a homing signal as I barreled toward Chapel Hill. The Tar Heels were in the process of battering Vermont, a game they would eventually win, 93 to 65. I swooped onto 15-501 and passed the exit for Duke. Soon I was home.

When I came into the house, my mother was listening to the postgame show on the radio. "Well, some days you are the pigeon and some days you are the statue," Tom Brennan, the Vermont coach, was saying. "Nobody has done that to us in two years."

"That's the kind of game I like," my mother said. "Not too close."

We sat into the night, watching an old detective drama on TV. "You don't see Perry Mason much anymore," she said.

"That could be because he's dead," I said.

She shot me a look that may have said, "Welcome home." Then again, it may not have said that.

I was tired. It felt good to lean back on the couch in my mother's house, my childhood home, the lulling voice of the dead Raymond Burr droning on about some wicked socialite with a talent for murder. My mother and I were going to spend a season watching basketball together. How many mothers and sons got to do that?

As I reclined, I realized that, like the socialite, I presented two faces to the world: one that of a quiet fellow who liked to read and walk and ponder things, and the other that of an absolute beast, a guy who screamed and ranted and jumped up and lay down on the floor as he watched North Carolina play basketball—a man who even at his advanced age hated to lose at anything. I decided that this season would be a good opportunity for the journalist in me to study the beast in me. That's what the journalist thought, anyway. There was always the possibility that the beast would decide to kick the journalist's prying ass right out of his own book.

That would serve that nosy motherfucker right, the beast in me said.

You really shouldn't use that sort of language, the journalist told the beast. It's not the sort of language a journalist uses.

Fuck you, you wimp, the beast responded.

I can see this is going to be something of a trial, the journalist sighed. You're going to get in the way of your own project.

My project is to enjoy things and indulge myself, the beast in me said. I don't ask you to change who you are. And I don't need you to change who I am. I've been this way my whole life. What's so bad about hatred, anyway? Isn't it just a human emotion that helps us determine how we want to live by identifying the worst alternatives? An animating faculty of passionate moral discrimination?

Your japery is only in the service of wickedness, the journalist said. And yet, it might make some people think you have the potential to be good. Or at least reasonable. But having lived with you for so long, I know better. I see deceit where others might get swept along by the puffy force of your self-infatuated rhetoric.

Always thinking the worst of me, the beast said. Why don't you run along and visit one of your interview subjects? And hand me the remote control before you split, if you don't mind.

"Will, you're talking to yourself," my mother said. "Why don't you go to bed? You've had a long drive."

"What was I saying?"

"Something about the animating faculty of moral discrimination. Next time, I'll take notes."

"Sorry."

"That's all right. Welcome home."

THE REMNANTS OF AN ANCIENT SEA

It had been nearly 25 years since I'd last lived in North Carolina. On every visit, I groused about the metastasizing development of the state, now the tenth or eleventh most populous in the country. The place seemed on the verge of becoming a giant suburb all the way from Murphy to Manteo. Where was this place?

My hometown of Chapel Hill had long served me as an anchorage of spirit, a phrase that Henry James had used to describe his grandmother's house in New York. I usually returned twice a year, at Christmas and for a few days in August. During both seasons, the town was mostly emptied of its student population. The streets were quieter, the stores less attended, and it was possible to imagine Chapel Hill as it had been when I was much younger and it was still called a village, though even then the description was a comical and tender attempt to preserve in name alone an earlier aspect of the place—its collegiate intimacy under the great thrall of old Southern trees. Professors bumping into each other at the downtown post office, for instance, all of them thinking their great thoughts, as my father, himself a professor at the medical school, used to say mockingly. He was one of them, and then again, he wasn't. But despite the town's genteel love affair with itself, its self-satisfied commemoration of its uniqueness, there was a gravitational power to its easefulness.

Chapel Hill sat in the Piedmont region of the state, almost halfway between the mountains and the coast, close to the specific juncture where the hill-mottled landscape gave way to the Coastal plain, which unrolled

as flat as a carpet from near Raleigh to the Atlantic. In a sense, the geologic facts embodied the way many North Carolinians felt about the
Chapel Hill professoriate—that it looked down its spectacle-crested nose
on the rest of the state, particularly in gazing eastward at Raleigh and the
university there, North Carolina State, an institution that began as (and
remained) a school emphasizing agricultural and technical education.

An actual view went with this mental projection. It was best afforded by
a stone bench at Piney Prospect on a grand hill just east of campus, under
the anachronistic shadows of Gimghoul Castle. As a student, I would
sometimes take a date there and tell her what my father had told me.
Pointing eastward at the flat pine expanse that led to Raleigh, I'd explain
how what we were looking at was the ancient bed of a Triassic sea. The
waves of that unnamed ocean washed right up against the base of this rise,
I said. In effect, we were sitting on a bygone beach! Wasn't that amazing?
The girl listened politely as I yammered on, increasingly excited by the
incomprehensibly vast scale of geologic time, the creaking degrees by
which the earth could change from one thing to another. Eventually, I
shut up and let silence reign—a better tactic. My date and I would simply
sit gazing out at this landscape that I loved, this epoch's swelling waves of
oaks and pine and hickories, tulip poplars and dogwoods, an ocean of
trees to the horizon, though increasingly dotted with the erupting volca-
noes of commerce. I realized that I had a lot of my father in me, especially
when it came to courting a gal. He presented my mother with his love for
North Carolina as if he were giving her a diamond necklace that would
always remind her of him, which, in fact, it would, it did.

This, of course, is a tale of two cities. Eight miles away stood Durham,
the city that did not so much contain Duke University as occupy a
proximate position. My mother used to take us children to downtown
Durham to the Young Men's Shop to buy our Sunday clothes. We went
to Sears to purchase our appliances, televisions and record players and
washing machines. In those days, there were no Best Products on the
road between Chapel Hill and Durham, no Circuit Cities or Office
Depots. Back then, the sweet smell of tobacco hung over Durham like a
loyal cloud, infiltrating every nook and cranny.

Duke was the monocle on the otherwise plain country face of Durham. Without Durham, there would have been no Duke. It was in Durham that the cigarettes were manufactured that were sold all over the world that made a fortune for the Duke family that endowed the college then known as Trinity, and that in 1926 changed its name to Duke University in order to honor its family of patrons.

On the campus, outside of the Duke Chapel, stood a statue of the school's prime benefactor, the tobacco tycoon James B. Duke, son of the patriarch Washington Duke, who once said of his son, known as Buck, "There are only three things in this life I don't understand—electricity, the Holy Ghost, and my son Buck." Incomprehensible perhaps, but Buck was a marketing genius who came up with the idea of including cards of pretty women in packages of cigarettes. He drove hard bargains, and eventually ruled a lucrative monopoly.

With evident satisfaction, his statue brandished a fat cigar between his burly bronze fingers. Smoking by students and teachers had been banned in the campus buildings, but a statue, at least, was still allowed to enjoy a good cigar after dinner. James B. Duke had endowed the school through the entrepreneurial vigor with which he sold his demon product. In essence, Duke was a neo-Gothic city of stone floating on a vast cloud of cigarette smoke. And now that university preferred to distance itself from its very origins.

The city of Durham was not so afraid of the besmirchments of commerce. It had arisen for no other reason. Just a railroad depot in the 1850s, the place had boomed as a manufacturing center after the Civil War when tobacco in the form of cigarettes began to be sold around the world. The factories and the warehouses teemed with workers, and a bumptious culture grew up around them. As the shifts ended, the laborers emerged to find bluesmen such as Blind Boy Fuller and Sonny Terry and Brownie McGhee playing for spare change.

Durham was a black and white town, each race taking neighborhoods of its own, all of them originally country people, sharecroppers and small farmers who had moved to the city in search of steady work. The neighborhoods reflected that; it was as if the country folk had carried their garden plots with them on their backs, stuffed into tote sacks and trunks.

Their unruly yards sprouted pecan trees and crape myrtles. They cultivated turnip patches, beds of tomatoes and yellow squash, okra and peppers.

The new city dwellers' rhythms were still country-slow, their summers still dictated by the oppressive air. At night, the porches swelled with men and women seeking relief from the heat incarcerated within their houses. The neighborhoods reverberated with low talk and laughter, the squeak of porch swings, the star-twinkle of cigarettes.

The famous black neighborhood of Hayti, a world unto itself, vibrated to its own set of rules. On the grander blocks, the sons and daughters of the black middle class studied classical piano. An insurance company called North Carolina Mutual, for many years the largest black business in America, had its roots here.

Durham bore about the same relation to Duke as a boiler room to a penthouse. The residents of Durham, black and white, shoveled in the coal, kept the boilers firing. And a few miles to the east, the students and professors at Duke turned the pages of the classics in the warm stone rooms of the university.

In 1881, shortly before the installation of cigarette rolling machines in 1884—an epochal moment in the tobacco business—the Dukes had lured into this traditional Southern world more than 100 cigarette rollers, Jewish immigrants from Eastern Europe by way of New York City. With the advent of the machines, the rollers were stranded in a culture that must have seemed unimaginable. Where they went, if they went anywhere, I had no idea. They had washed up in a strange land, I suspected. But I was home.

DEVIL WORSHIP

My first morning at home, my mother and I were inspecting Christmas trees in the parking lot outside of Whole Foods, the archetypal university town grocery, with its Sixties fetish for the organic and its contemporary habit of charging too much for tenderly coddled vegetables and roots. The black guy selling the trees wore a Duke cap. It turned out that he was a divinity student at Duke. I asked him how he could wear a hat like that in this town.

"Duke is number one," he said to me, holding up his index finger.

I thought of holding up a finger of my own, but I didn't. Instead, I said to the Duke divinity student: "And you, a man of God."

"Worshipping the blue devil," my mother chimed in.

The divinity student stayed on message. "Go Blue Devils," he said.

We bought a tree from him anyway. Driving home, I said, "We double-teamed him."

My mother said, "Do you think he sold us a dried-out tree?"

"It seemed okay," I said.

But lo and behold, that tree began to shed its needles as soon as we put it up. And no amount of water and molasses could put it right.

My mother came to basketball when she was firmly settled into middle age. She had grown up in Springfield, Massachusetts—birthplace of the game, home of the Hall of Fame; were these not omens?—where she played field hockey, went off to Mount Holyoke, studied chemistry, balanced books on her head to test her posture. She came to basketball through her husband and her children—especially her children—and she fell for the game with the ferocity of a late convert.

Now she watched contests in a state of crackling high tension, as if plugged in to a wall socket, or, actually, a power pylon. She leaned forward toward the television, her fists clenched to her chest, brow furrowed, eyes locked to the action. Because it was her nature, she tended toward pessimism as a protective measure, an offering that might forestall greater doom. She magnified the reversals suffered by the Tar Heels, underplayed the successes. "Oh, no; oh, no," she groaned when she sensed that one of the players was about to screw up. If Rashad McCants, say, clanged a three, if Raymond Felton lost the ball, if Jackie Manuel even thought about an outside shot, she squeezed her fists and mourned prematurely. "They're going to lose. I know it. I don't even want to watch."

Indeed, at such moments, she often disappeared into the kitchen and found solace among the clattering pots and pans, the soothing rumble of the dishwasher. And if the Heels really did lose, she was never one to listen to the postgame analysis. What was there to analyze? They'd lost. She'd snap off the television or the radio with crestfallen finality.

If only victory had been a sufficient salve for the losses. But my mother never seemed as happy from a win as she was distressed from a defeat. Winning had come to be a baseline requirement in the same way that for junkies, heroin eventually becomes necessary not for the ecstasy it originally provided but for the routine maintenance that prevents the sniffles. Or the doom of withdrawal. It was sort of like that for my mother.

Often after wins, she announced: "That was pretty good. They looked all right."

"*Pretty good? All right?*" I'd roar over the phone from New York as we discussed a game just finished. It was my custom to phone immediately after games to replay moments, consider strategy, and evaluate which way the team was tending. "That was better than *pretty good.*" In fact, my mother engaged with games more like a coach, wary and anxious about what minor deficiencies portended. When my father answered the phone on those game nights, he would say, with resignation in his voice, "Let me pass you on to the coach," and then fall back on the couch to continue his snooze. Defeat, victory: Neither was reason enough for my father to miss out on a good winter's-night nap.

I also suspected that the "pretty good" gambit was a keen psychological ploy for minimizing future loss. But the 8-and-20 season the Tar Heels had suffered in 2001–2002 had taught me to celebrate even the smallest moments, because everything could be taken away. My relation to the team's success these days was that of a perennially broke guy who got a check in a legal settlement and went out that afternoon and bought himself a Cadillac with all of the options. One dunk and I partied like it was 1993.

My mother, by contrast, was sneaky. She tried to placate the New England God of her youth, that stern Congregationalist who frowned upon too much celebration. She tried to make a deal wherein if she wasn't too happy with the wins, they would keep on coming. Stay low, stay inconspicuous, and the gods—the God—may not single you out for punishment. Maybe Raymond Felton's hot streak of threes will continue. Maybe Sean May will keep the weight off. My God whoop-whooped. Her God pooh-pahed.

On those rare occasions when the Tar Heels were losing, my father liked to get under my mother's skin by asking, "Gloria, how much do you think they're going to win by?" First of all, he was well aware of the team's reputation for miraculous comebacks under Dean Smith. So he might have been trying to teach her to have a little faith. But knowing my father, I suspect his intent was different. If you want to torture someone who desperately wants a win, guarantee her that win. Goad her. Raise that win in front of her like a red cape. It can be infuriating. How was he able to remain so detached from the fray? She cared. He didn't. At least, not as much. Unless it was Duke.

THE RELIGION OF FOREHEAD

Raising his children in the International Brotherhood of Duke Haters was the natural and one of the more enjoyable aspects of my father's master plan, though "plan" is probably too intentional a word for the improvisations of child rearing. His intent, let us say, was to develop his children into good Southerners of the North Carolinian persuasion. He wanted us to embody the values of our family, going back generations. To an observer, a foreigner in our midst (say, like me), it may have appeared that my father intended to turn us into younger versions of himself.

But there were obstacles. My siblings and I were growing up in Chapel Hill, a university town in which cosmopolitan toxins pervaded the air like the odor of dogwood blossoms. Or was it the scent of Thai stick? Ours was the town that Senator Jesse Helms had suggested surrounding with a chain-link fence in order to build the state zoo. Or so we liked to brag. It wasn't necessary to go to New England every summer and visit my mother's family to absorb dangerous diction, hard consonants, and mean vowels. That could come to us every day from our teachers and coaches and friends and their families, and from that demonic television set in the breakfast room that spoke a liltless tongue called Middle American.

And so began our supper-table lessons in how to talk right. Notice that I said "supper," not "dinner." North Carolinians of my father's generation did not sit down to dinner at night. They sat down to supper. Dinner took place in the middle of the day.

"How do you say "F-O-R-E-H-E-A-D?" my father asked us many a supper, spelling out the word lest he give away the answer. He and my mother sat at opposite ends of the kitchen table with the four children between them, my sister, Annie, and I on one side, my brothers, John and David, on the other. My mother listened to the interrogation with bemused and anxious forbearance. My father monitored our answers like a detective taking a suspect's statement.

"F-O-R-E-H-E-A-D. Come on. How do you say it?"

Most Americans these days would answer *for-head*, rhyming the word with *oar-head*. But that was not the answer my father was seeking. That was wrong. Only ignorant barbarians said *for-head*. Or people from other states, who probably had no choice but to be ignorant barbarians. We had a choice.

Sooner or later, someone (the younger siblings usually succumbed first) would chime in with the correct response, which was *far-red*, the first syllable rhyming with *tar*, which was appropriate for young Tar Heels like ourselves. To answer otherwise, as I sometimes did out of sheer perversity and because in fact my father was teaching us how to stand our ground in the larger world (I just applied the lesson at home, too), was to witness him shake his head mournfully and tell us (me) that we just had to rebel, a tendency which must have come from my mother's side. Why couldn't we just accept the truth of certain things, like the proper way to say F-O-R-E-H-E-A-D, instead of disputing or making fun of them?

I sometimes ventured that a word might be pronounced one way in one part of the country and differently in another. He would tell me that in the case of F-O-R-E-H-E-A-D I was wrong. And that even if I had been right, it was hurtful to the discriminating ear to hear *for-head* rather than *far-red*.

"How do you pronounce F-L-O-R-I-D-A?" he asked.

"*Flar-da.*" Resistance was useless.

"F-O-R-E-S-T?"

"*Far-rest.*"

"That's the ticket! Hotcha!" Oh, how happy he was when we gave him the pronunciation for which he yearned. He beamed with family happiness. He ushered us warmly into the exclusive club of good North Carolinians, true Southerners.

"You all seem so close," more than one person has said about my siblings and me.

"We had a common oppressor," I tell them.

And yet, drilled though we were, watched over like prisoners in the yard, hectored and pursued like lovers who might spurn their suitor for another (for that is what this was really all about), I understand now with great sympathy what my tyrannical father was after. He wanted his children to be able to speak with the ghosts of our ancestors, to preserve through language a realm outside of time. He desired that we speak that ancient tongue to him. And in doing things in the old way, we would link ourselves to family members both dead and gone and yet to come. Saying "forehead" was like lighting a candle in our religion.

And so was hating Duke. It was part of who we were, how we defined ourselves in a world unmoored.

THE BETRAYAL

On Christmas Day, 2004, we gathered around our balding tree, opened presents, and remembered Christmases past. My father's absence was still palpable; we missed his joy in the yearly rituals, his fireside sermons on the religion of family and home. He had regarded my life in New York as a betrayal, one forced perhaps by circumstances, but a treachery nonetheless. In his eyes, my domicile made my credentials as a Southerner somewhat suspect. All I had to do was eat supper at eight o'clock—"New York hours," he grumbled—and I was a turncoat. A true Southerner ate supper at 6:30, which, as it happened, was when my father liked to eat.

One Christmas in the late Nineties, not long before he died, I sat across from him in the living room, both of us arrayed in the same positions we'd taken 25 years before, during The Years of Teenage Rebellion. I can't recall why (maybe it was the punch we were consuming, which contained four different types of liquor), but for some reason, I started enumerating the things I missed about North Carolina. There wasn't anything exceptional about my list. It included such Southern staples as the sound of june bugs on a blistering July afternoon; the politeness to be had at the 7-Eleven, where the manager was always ready to pass the time of day with you at the expense of speedy service; and the slow-cooked wisdom of the state's old liberal avatars, such as Judge Dickson Phillips, a friend of my father's whom we both admired. Judge Phillips had once cut through my angst at deciding where to go to college by asking me a simple question: "Will, are you a happy person?" I suspected that I might be, and I told him so. "Well, then, you'll be happy wherever you go," the judge said. Case closed. Time to watch football.

My father listened quietly as I talked that night, and I thought nothing more of the conversation until the next morning, when he came to the bedroom where my wife and I were packing to head back to New York. He stood at the door, watching us and weeping.

My wife and I looked up from the suitcase we were at that moment coincidentally stuffing with MoonPies, a nutritious Southern staple made of marshmallows, chocolate, and all sorts of delicious hydrogenated grease that I liked to bestow upon my friends and coworkers in New York. The easiest place for me to be a Southerner, I had discovered, was in Manhattan. Give a Yankee a MoonPie and they look at me like I am Robert E. Lee. Or Hank Williams. Or Bear Bryant. "What is it?" they ask.

"Why, chile, that's a MoonPie," I say.

"What's in it?"

"I can't tell you what's in it. But I can tell you that it'll put hair on your chest and lead in your pencil."

So here we were, packing our MoonPie contraband into our luggage, and here was my father at the door, crying. "I wish all of you could live down here," he said. He had frequently offered us a patch of land behind the house, a little perch on the hillside, where he hoped we would build a residence and establish a compound of Blythes, all within hollering distance of one another.

"I wish we could, too," I said.

"That meant a lot to me, what you said about North Carolina last night," he said. "I didn't know you felt that way."

"Well, I do," I said.

"Why don't you move back, then?" he asked.

"Can't right now. Work and all. But I'd like to."

"I don't know about that. I doubt you're ever coming back," my father said. Now he was returning to another one of his roles: skeptical victim of his children's incomprehensible decisions. The King Lear of 114 Hillcrest Circle. He turned to trudge down the hallway, the tears still pearled on his cheeks, his frame looking suddenly smaller and swallowed-up in his khaki pants and his worn white shirt with its sleeves rolled up to his elbows.

THE FAMILY THAT HATES TOGETHER

A family that plays together is a family that stays together, but a family that hates together is a family that really loves each other, everyone glopped together like a ball of sticky rice. Let me show you what I mean. It is game night this season at the Blythe household. Through the wind-swayed winter woods, the lights of University Mall flit and twinkle in the distance.

Inside, Sheba the dog slumbers on the rug in the TV room. But as tip-off approaches, she hauls herself into the kitchen in anticipation of the tumult to come. Like many animal species, dogs are said to have a sixth sense when it comes to impending earthquakes.

There on the sofa to the right of the TV is my mother, silver-haired and dignified, more kindhearted than anyone I have ever known. Her bills and church-circle correspondence are stacked beside her, and if North Carolina's lead is a comfortable one (for her, around 40 or 50 points), she might even get a little work done. She is a good Presbyterian woman. And she hates Duke.

On the big sofa across the rug from my mother, you will find my sister, the reporter. Objectivity is her middle name. She lectures me about being faithful to facts. She is concerned that I might play too loose with them. I tell her she is a real journalist and she is kind to be concerned but that I am something else. When she's not writing her objective accounts of local citizens, my sister lives the genteel life of a gardener, tending her flowers and trees, worrying over her peppers and tomatoes. She might say about herself that she is naturally shy but that she has learned how to talk to people, which indeed she has. She cooks an occasional meal for my mother and me, and in my time in New York has become an extraordinary cook. How did this happen? Soon, my shy, pepper-growing sister will unleash upon the television set a harangue that would brush back Don Rickles. She, too, hates Duke.

I sit next to my sister on the big couch, though "sitting" is an imprecise term for what I actually do. For me, the next couple of hours will be all about the positioning, about the spin I can put on the game by contorting my body into necessary postures. Really, it all depends on the flow of the game. As coaches like to say, you shouldn't decide all of your tactics ahead of time. You need to remain sensitive to every shift in the action. So, as you can see, I am a reasonable man. I try to do right in this world. But I, too, hate Duke.

Were my brothers, David and John, available, they, too, would cluster here, one on the rug, one parked next to my mother. Wives, girlfriends, children moving in and out of the room: They knew that this was not their fight and that it was hopeless to try and impose normal standards of etiquette on the proceedings.

One family, united in the dark sacrament of disdain, facing the world together, side by side, couch by couch. We've got each other's backs and we're ready for the game to begin.

THE BEAST IS ALIVE
Duke versus Clemson
January 2, 2005
Cameron Indoor Stadium

Tonight we are watching Duke play Clemson. We are monitoring our adversary for cracks, structural defects, familial dysfunction. We want to know who mopes, who snaps under pressure, who misses. We are scholars of the slippery slope, the January weaknesses that portend doom in March—the point guard who can't shoot, the two guard who can't defend, the center who clanks free throws.

The new year has arrived but when it comes to hatred, the beast is already in midseason form. He lives not just in me, his favorite host, but in my mother and my sister.

"How can anyone stand to look at him?" my mother asks, staring at Mike Krzyzewski.

"It's a mystery to me," my sister says. "One that is simply beyond our human capacity to understand."

"My friend Nina Wallace can read lips," my mother says. "She says you ought to *see* the kinds of things he says."

These are the kind of things we say. The game is ugly, too. At one point, the commentator compares it to a root canal. Both teams are building a brick wall of missed shots. With 7:34 to play in the first half, the score is deadlocked at an unimpressive 10 to 10.

The second half is more of the same: turnovers, fouls, neither team shooting over 35 percent. "Why do they get to foul so much?" my sister asks of the Blue Devils.

"Because they're Duke," my mother says.

"That's another one of those profound mysteries," I tell my sister.

"Sit down," my mother instructs Mike Krzyzewski.

"Miss!" I shout at Lee Melchionni. His three-pointer ripples through the net and he runs down the court, fists clenched, screaming.

"They're so lucky," my sister says.

"I hate the way he screams after every shot," I say.

Clemson makes a game of it, pounding the boards. The Tigers actually take a 38-to-35 lead midway through the second half. Our misanthropic band cheers. "Whip their sorry asses," I say.

My mother appears to suppress a smile. She doesn't ordinarily like that kind of language, but there is a time and a place for everything. "Come on, fellas," she says to the Tigers, which I think is her way of saying something similar.

"Wouldn't it be great if they won?" my sister says.

But they don't. JJ Redick scores 20 of his 24 points in the second half. He's not bashful about putting it up, even though he's only four of 11 from three, eight of 19 from the field overall.

"They don't look that good," I say.

"They're beatable," my sister says.

"I hope you're right," my mother says. "Because I can't stand them." That's our family. Sweet right up until tip-off.

I remember the spring afternoon my girlfriend watched me in horror as I ventilated my darker passions while watching Duke play Carolina. I sought to quell her anxiety. "I'm having fun," I told her.

"That's what that is?" she asked.

"Yes, most assuredly," I said.

"I don't think I'd want to see you not having fun," she said.

"But you have," I said. "It looks sort of like this but different."

"Right," she said.

"The key thing is not to take it personally," I reassured her.

"I don't."

"Good. You shouldn't."

"Your whole family," she said. "They seem so civilized. So nice."

"They are," I said. "They're really nice."

"But when you guys watch basketball . . ."

"Yeah, I know. That's when the beast comes out."

Listen, I'm not really justifying this. I'm explaining, not excusing. I mean, there's a reason or two, or maybe a couple of dozen, why I am this way.

Mostly, I'm well-behaved as a journalist ought to be. Mostly, I am studying myself and those around me from a cool, interstellar distance. It's curious how this Duke-hating and Carolina-loving became so intense.

But every now and then, I can't help it. The old impulses reemerge. I watch a game and I go bonkers. It seems the whole universe is tied in to the game. But I worry that by such deep immersion in this obsession over the next few months, I may start to dissolve it. That can happen—it's the smoke-ten-packs-of-cigarettes-the-day-before-you-quit theory of killing an addiction. I can't quite imagine life without basketball, however.

I told this to a friend and she said, "Oh, that's good. Maybe this will cure you somehow."

"But I don't want to be cured," I said.

THE PLEASURES OF HATRED

At times it troubles me a little to be so full of piss and vinegar. A man of my age ought to be seasoning into acceptance like a salt-cured ham. The study I've done of Buddhist literature (such study being much easier than the actual practice of Buddhism) suggests that not only is hatred bad on a cosmic level, but it is also bad for us personally. We are going to pay for our bad deeds and our evil thoughts by being reincarnated again and again, swirled from one life to another like dirty clothes on endless spin cycle. Although it isn't very Buddhist of me to worry about ending in the karmic washer for a few billion extra millennia merely because *I* have the odd hateful thought—I should be more concerned with others ending up in such straits—I can't help but fret about my fate. One North Carolina–Duke game alone probably costs me several millennia of rebirths.

And yet, how I hate.

From across the centuries, I recently found good company in the English essayist William Hazlitt, who died back in 1830, nearly friendless, it is true, and with hardly a tuppence in his pocket. But no one ever said hatred is the best way of winning a man friends and money. Samuel Coleridge, one of Hazlitt's erstwhile friends, described him as "ninety-nine in a hundred singularly repulsive." Hazlitt's wife left him on returning from their honeymoon. And yet as he lay dying in a tiny room, Hazlitt said, "Well, I've had a happy life."

Could hatred, like prayer or Prozac, have been the secret? In 1826, Hazlitt wrote an essay called "On the Pleasure of Hating," a profound work that expresses what we might call a holistic view of hatred. He puts the noblest face possible on a snarl. "Nature seems made up of antipathies," he proposes. "Without something to hate, we should lose the very spring of thought and action." This struck me as quite likely—that every bit as much as love, hatred moved a person to ponder and to act. And that such hatred needn't even be personal. It might be disdain for injustice, for poverty, for drunk driving. Hate the sin, not the sinner. Or so it seemed upon first reading. Things actually got more complicated.

Hazlitt had detected the boredom inherent in goodness, the totalitarian features of the standard-issue heaven. "Pure good soon grows insipid, wants variety and spirit," he writes. "Pain is a bittersweet, which never surfeits. Love turns, with a little indulgence, to indifference or disgust: hatred alone is immortal." I knew marriages like that, kept alive, if not flourishing, by endless campaigns of attack and counterattack, animosity an apparently inexhaustible fuel of togetherness.

Hazlitt's final estimation of the dynamics of hatred, "the wild beast," proves fascinating. Hatred alights on conditions such as injustice to express itself. Hazlitt takes a swipe at religion as one of the prime venues for this basic human need for antagonism. "What have the different sects, creeds, doctrines in religion been but so many pretexts set up for men to wrangle, to quarrel, to tear one another in pieces . . . ?"

The author expresses himself in that English manner we've come to think of as commonsensical. "Public nuisances," he writes, "are in the nature of public benefits." That is, they not only excite the body politic, they sting it into collectivity. "How long did the Pope, the Bourbons, and the Inquisition keep the people of England in breath, and supply them with nicknames to vent their spleen upon!" Had they done us any harm of lately? No: but we have always a quantity of superfluous bile upon the stomach, and we wanted an object to let it out upon."

Yes! I know just how that misanthrope felt. Substitute Duke for the Pope, et al., and, well, Duke had sopped up a lot of superfluous bile in my day. In the early Nineties, for instance, I seemed to have hated Bobby Hurley, the short and skinny Duke point guard. Yes, frail Bobby Hurley, whom by all rights I should have identified with for his striving against the odds, his fearless adventures in the lane among the giants. Bobby Hurley who cried when North Carolina point guard King Rice muscled him around his freshman season.

Hurley even made certain American virtues suspect because he was the one embodying them. Grit, for instance. Hurley definitely had grit. You didn't need Dick Vitale to point it out a thousand times. You could see it yourself, watching this New Jersey white kid blaze up and down the court like a Chevy Camaro about to throw a rod. He drove the lane with the bony éclat of an overachiever unafraid of a pounding. And at least he cared enough to cry.

Let me be the first to admit that had Bobby Hurley played for UNC, as he had wanted to do before Dean Smith told him he would have to wait for high school star Kenny Anderson to decide where he wanted to attend college, I might have seen those defects for the virtues they can be. Hatred, therefore, teaches us that context is everything. Our passions are fickle; they'll enlist in whatever battle we care to fight.

I have come to believe that worse than an honest hate is a dishonest piety. How much further from heaven is such a posture, despite the sack-cloth (or khakis) that such folks wear. And it seems that North Carolina is crawling with such pious types.

THE NORTH CAROLINIANS

We are modest and immoderately proud of that modesty, a somewhat paradoxical condition. We like to cite with self-effacing pride the famous quote: "North Carolina is a vale of humility between two mountains of conceit." Those mountains, of course, are the old aristocratic bastions of Virginia and South Carolina, though now they too are becoming chopped-up subdivisions of suburban values similar to North Carolina's. The aristocrats have all gone underground, as in *died*.

The North Carolinian believes in niceness as both a practice and a veil, as an expression of Christianity and democratic values, and as a disguise. Even the Carolinian's most unseemly moves will occur behind a screen of Christian rectitude. Or a silly Christmas sweater that suggests roly-poly affability. Again, a prideful possession of the middle.

The average male North Carolinian wears khaki pants the way other men wear fancy suits, as a badge of his identity. He won't get too far above his raising. The choice these days is really between pleats or flat-fronts.

We are what we are.

Unable to express hostility with much directness, restrained as we are by an ill-fitting Christianity (and if male, those middle-of-the-road khakis), we North Carolinians will come at you sideways with either shy irony or truculent sweetness. First, we'll try to save you from yourself. If that doesn't work, we might try to kill you with kindness. And if that doesn't work, we'll just kill you, but politely. This is what good manners are for. We're masters of a complex, coded set of manners. There are many ways to kill that don't require actual bloodshed.

I say that the North Carolinian is unable to express hostility with much directness—that is, unless he or she has the mixed blessing of being a redneck, very few of which still exist in their original state, having been suburbanized to the point of neuter by improved economic circumstances and Republican campaign strategists. Indeed, the cult of the redneck, as celebrated in popular culture by the likes of the late newspaper columnist Lewis Grizzard (who never met a bowl of grits he couldn't eulogize like Pickett's Charge) and by the comedians Jeff Foxworthy and Larry the Cable Guy, is a sure sign that the redneck's original badass vitality has been largely sucked dry. Toothless and enfeebled, and in that sense looking a lot like he did in his prime, the redneck has been helped into the Museum of Old Stereotypes, an institution that has been franchised and broadened to include NASCAR racetracks. At the big raceways like *Talladega* and Rockingham, an entire species stuck in limbo between old country ways and the tedious demands of suburbia celebrates the wild way life used to feel, even if they never felt it personally. Remember them bootleggers cornering on a hairpin mountain turn? The cars roar by, and the fans shake with the engine vibrations as if they were equipped with all that horsepower themselves. Want to see rednecks these days? They're safely sealed inside television glass. They're a concept, not a class. They're caricatures of disenfranchisement sold by intellectual merchants who've found profit in resentment.

The suburban Carolinians have an entirely different relationship with the world. They've been civilized. Where their redneck antecedents may have cursed and grumbled and taken the Lord's name in vain (or shouted for Him in desperation at a late-summer revival), the suburbanites slather prayer over everything like ketchup. They believe in a personal relationship with Jesus that makes the Lord sound like the best bank loan officer/therapist/career counselor a guy or gal could ever have.

The new Carolinians like to preface the word "family" with the definitive article "the," as if family were something new and brave in the

world, a political party of its own. The way they say "the family" makes
the two words reverberate together with Mafioso undertones and suggests an exclusive organization prepared to eliminate all competitors.

Such Carolinians inhabit their suburbs like little islands unto themselves, each family a separate nation. The Reconstructionist ideal might
have been 40 acres and a mule. These days, the Carolinians want 1.4
acres and a gleaming SUV that's as durable as a mule and a whole lot
shinier.

So much niceness! Such family values! Between megachurch and
maneuvering decently at the office, how's a fellow or a gal supposed to
cut loose? What if a bad thought crosses that sweet, Jesusified mind?
Where's it supposed to go?

Well, brothers and sisters, that's where basketball comes in. In a state
otherwise deprived of outlets for the vehement passions—that forbidden emotion of hatred, for instance—there is ACC basketball, a veritable
festival of hatred, sanctioned and almost sanctified. And the most antag-
onistic relations within the ACC can be found between Duke and North
Carolina.

THREE

The Coach in the Basement

THE COACH IN THE BASEMENT

SOON AFTER I RETURNED TO CHAPEL HILL, I had arranged to visit with Dean Smith, the former North Carolina head coach. I dressed up for the occasion, although these days the journalist in me had become so blasé about interviews that he tended to walk out of the house wearing whatever happened to be hanging on the nearest chair. But for Smith, I chose a dark, pinstriped suit, the sort you'd wear for an audience with the Pope, if you were the kind of guy who actually wanted to see a Pope.

As I drove across town to the Smith Center on a bright and shining winter noon, it was clear enough that the world hadn't ended at 3:00 P.M. on Thursday, October 9, 1997, but it had sure felt that way to me at the time. That was the moment when Smith stepped down after 36 years as coach of the University of North Carolina's men's basketball team. As a Tar Heel expatriate doing hard time in Manhattan, I'd dreaded the day for years, and when Smith announced his retirement, it felt awful, worse even than losing to Duke, though for the record I am compelled to note that that had happened only once in Smith's last nine games against the Blue Devils. Since then . . . well, let's not talk about it.

Phil Ford, the former All-American point guard and then an assistant coach for the Tar Heels, captured the shock and melancholy of Smith's departure when he said that it was "like dying." He was right. In fact, I'd lost relatives with less disruption to my world.

Dean Smith made basketball a religion in our family, rivaled only by Presbyterianism, barbecue, and the Democratic Party. For many years, our dream scenario had gone something like this: Dean beats archrival Duke and its coach, the detested Mike Krzyzewski, for the national championship in April, then runs for the Senate (for which he had long been rumored a candidate) and whips the antediluvian Jesse Helms in November! Then we all eat barbecue from Allen & Sons, on Highway 86, to celebrate. Then we go to church to thank God for sending us this wry Kansan to make our state safe for hoops and liberals. Sadly, this never happened.

By the time he retired, he'd won 879 games, more than any other coach in NCAA Division I history. Since taking over the head spot from the charismatic Frank McGuire in the 1961–1962 season, he led Carolina to two national championships, 11 Final Fours, and 13 ACC Tournament championships. Under Smith, the Tar Heels won at least 20 games for each of 27 straight seasons, 30 of his final 31. In 1976, he coached the Olympic team that restored the gold medal to the United States. In 1983, he was inducted into the Naismith Memorial Basketball Hall of Fame. He had been selected by ESPN as one of the top five coaches of the twentieth century in any sport. In 1998, he received the Arthur Ashe award for courage. And his teams had graduated more than 96 percent of their lettermen. With Smith, we got to applaud not just victories, but a steadfast moral vision. It might have made us, his admirers, a little insufferable from time to time.

Over the years, we had studied him intensely. He was something of a stern mystery, a Midwestern preacher-coach thriving in the honeysuckled extravagance of the South, a play-it-close-to-the-vest Kansan whose propriety and saintly soul never quite concealed the capabilities of an assassin. He put sportswriters to sleep with his monotonic poor-mouthing (yes, we, too, respected St. Mary's of the Blind, but we knew that on any given night they couldn't really beat us, despite Dean's caution). He could be something of a pious sandbagger. But you listened to him anyway, for the occasional truth he couldn't quite hold back, like when he announced to the world that Duke's two white stars from the late Eighties and early Nineties, Danny Ferry and Christian Laettner, had lower combined SAT scores than those of J.R. Reid and Scott Williams, two black players who were often the brunt of jokes by the Cameron Crazies; "J.R. 'Can't' Reid," they chanted during one game.

It wasn't that we didn't have our criticisms of El Deano (as we affectionately called him, though not to his face). He could be obstinate. In the pre-shot-clock era, he resorted too soon to the four corners, his brilliant delay game, requiring kinesthetic marvels like James Worthy and Michael Jordan to stand around, holding the ball. To the day he retired, his teams appeared to defend the three-point shot as if it were still worth only two points. His hoarding of time-outs turned us into amateur psychologists, speculating on his upbringing. What had he been denied at an early age? He substituted too much, and showed faith in reserves who seemed astonished themselves to be out on the court at crunch time.

In 1993, near the end of the nerve-wracking championship game against Michigan in the New Orleans Superdome, Smith inserted several subs, including the little-used senior guard Scott Cherry. I remember rising in my seat, wondering in unison with thousands of other disbelieving Carolina fans, "What is Dean doing?"

But as usual, Dean's ploy worked. He got his starters some vital rest, and they played sensationally in the final seconds.

We could criticize him, of course, because we loved him. I wondered if we would ever see his like in college sports again. In fact, his type was disappearing from the world faster than the rain forests. He was that admirable paradox: the authoritarian liberal, the progressive who wasn't a moral relativist. He treated all of his players equally. He ran a clean program. He helped integrate Chapel Hill restaurants. He supported a nuclear freeze. And he practiced Christianity quietly without the rabid proselytizing so common among the bounty-hunting Christians of today. He emphasized the virtues of tolerance, persistence, and respect.

Sometimes, his rules seemed almost niggling; if a player was late for a team meal before a game, he was held out of the game for exactly the amount of time he was late. It was one of his many lessons in responsibility to your coaches and teammates. But Smith never forgot that responsibility was a two-way street. You weren't just responsible to him. He was responsible to you.

And unlike most present-day liberals—how strange that the label has become so disdained—El Deano won. He won a lot. Just ask that Republican Mike Krzyzewski.

It was ironic, I confess, given all of the above, that I had gone to visit Dean Smith to ask not only about Duke and North Carolina, but also about learning how to lose. But this was the sort of information that middle age demanded. I figured Smith had to know a little something about losing, or long ago he would have been poisoned from internal pressure like a diver with the bends. Admittedly, there was a paradox here; only the fact that Dean won so much underwrote his opinions about loss.

My trip to the mountain ended in a basement. For that is where I found Dean Smith—in the basement of the Dean E. Smith Center—the basement of his very own building. Ah, the contrasts were too easy to draw: Duke's Mike Krzyzewski on the top floor of his admission-by-scanner-only stone tower, the pine-needled light shining through the high windows; Dean Smith in his basement under a maze of pipes and ducts, a fluorescent world of cinder blocks and concrete floors, of dollies and spare seats and forklifts, in which he shared a windowless suite with his longtime assistant and brief successor, Bill Guthridge. On the wall of the outer office hung a Bruce Springsteen concert poster.

The cynics might call this self-effacement gaudy in its own way, conspicuous inconspicuousness. The true believers would regard the choice of location as consistent with Smith's nearly Quaker modesty, his aversion to the growing American predilection for self-promotion. He showed what he thought about the trappings of success by moving into a cave underneath his own dome.

Smith met me with a can of Diet Coke in his hand. In contrast to my tributary apparel, he was casually dressed. His hair was whiter, but the famous nose still splayed the air like an ice cutter slicing through the Northwest Passage. He had once received mail from a North Carolina State fan addressed simply to "Coach Nose, Chapel Hill."

Smith had a twinkling way of deflecting the expectations generated by his status of an icon. He asked me about people we knew in common, told me how much a friend of his had respected my father. He proved remarkably unassuming, but why would I have expected anything less of this particular man?

The passionate nature of the rivalries between the North Carolina schools surprised Smith when he first arrived at UNC. "Coming from the Air Force Academy," he said, "I didn't expect anything like this." He attributed the ferocity partly to the strong personalities of McGuire and North Carolina State's Everett Case, the Indiana native. McGuire was a flamboyant Irish-Catholic from New York who wore custom-made suits with handkerchiefs in their pockets, and who dined at posh restaurants without much concern for who paid. Taking the North Carolina job in 1952, he delighted the locals the way Jim Valvano would do one day at NC State. That is to say, he won, and he was a spectacle. Case was an Indiana native, a lifelong bachelor looked after by his maid. He arrived in Raleigh in 1946 and began to popularize basketball in his new state. Up to that point, the sport had been merely a bridge between football and baseball, something to occupy what was then a mostly agricultural region after the tobacco had been sold at auction and the farmers were lying low until spring. Case, too, won a lot. After his death, it was discovered that he had willed much of his estate to his former players.

"One thing people always misunderstood about Frank and Everett," Smith said, grinning. "Frank would always run over to shake hands. Everett didn't like to do that. He always ran to the dressing room. He said, 'We'll shake hands down below.' Frank said, 'No, I'm going to come after you and shake hands, otherwise you're running from me.'"

Like McGuire and Case, and like his rival, Mike Krzyzewski, Smith, too, had arrived in North Carolina from elsewhere. Born in 1931, he had grown up in Emporia, Kansas, then Topeka. It intrigued me how the prevailing sensibilities of Kansas and North Carolina seemed so closely matched. Each state valued the plainspoken, the modest, and the unadorned. Within that plain speech, of course, there were many ways for a man to shade his meaning, to say both more and less than was suggested by the face value of the words. Smith was a master of this. "I say what I think," he said, "just not everything I think."

If Kansas were a Southern state, it would have been North Carolina, and if North Carolina were a Midwestern state, it would have been Kansas. Both states staked out the middle as if that were the most honest place to be. Topographically, Kansas lay for the most part as flat as a pool table, and North Carolina enjoyed both mountains and a coastline, but in terms of cultural geography, they were brothers. I suspected that this might be the legacy of both states' agricultural roots. That such cultures might tend towards the laconic struck me as a reasonable response to the fact that crops were rarely impressed by the fast talkers of mercantile societies. Crops needed rain, not promises.

I tried this theory out on Smith's longtime assistant and fellow Kansan, Bill Guthridge. It rapidly became clear that he had no idea what I was talking about.

In Kansas, Smith enjoyed an All-American boyhood, the son of stern but loving schoolteacher parents. There were inklings, though, of a more complicated world. He recounted a story about himself in high school. He played quarterback on the Topeka team with a talented black receiver named Adrian King, who he remembered had once saved a game against Salina by adroitly snaring a bad pass of Smith's late in the fourth quarter, tying the score. "But they wouldn't let him play basketball," Smith said. "I went to the principal, because he was an ex–football coach who was a friend of Dad's. I said, 'You know, Adrian would really help our team. Why is it he has to play on his own team?' And the principal said, 'Dean, that was just something we decided to do because we'd have trouble with all these . . . you know, at the dances after the basketball games.' And how dumb I was, I said, 'Oh,' and walked out."

Despite his early run-ins with racial prejudice, Smith was still shocked when he came South the first time and saw a drinking fountain marked "Colored Only." In tandem with a young preacher fresh in town named

Robert Seymour, the leader of the new Binkley Baptist Church, Smith set about testing the willingness of local restaurants in Chapel Hill to conform to the new law guaranteeing all races equal access to businesses and public facilities.

"Chapel Hill got a reputation for liberalism from Mr. Howard Odum and the Sociology Department at the university," said Seymour, one of Smith's closest friends. "But Chapel Hill was as segregated as Mississippi. The only place in town where you could sit with a black was Danziger's, a place run by a refugee from Hitler's Germany or Austria. The former senator Frank Porter Graham said that Chapel Hill is like a lighthouse. It sends its beam into the distant darkness, but like a lighthouse, it is also dark at the base."

Smith and Seymour decided to take a black guest to the Pines Restaurant. "We were going to test them," Seymour said. "Dean had a vested interest in the Pines because the basketball team ate its pregame meals there. Everything went smoothly. The Pines, along with other restaurants, had been castigated by people in the community for resisting reform. We turned out to be like any other Southern town."

But it was in that town, my town, that Smith became a moral exemplar. I remember him in his big, pimpin' Carolina-blue Cadillac, an ocean liner of a car, nosing into a tight spot at Granville Towers during his annual summer basketball camp, at which I had managed to secure a spot through the intervention of our next-door neighbor, a donor to the athletic program. The Cadillac itself had been a gift of boosters.

Dreamer that I was, raised on sports biographies, with their morals of pluck and practice rewarded, I tried to catch Dean Smith's eye at camp. He himself had devoured the sports novels of John Tunis, having been required by his mother to read a book a week. So he knew a little about fantasies of persistence and discovery.

We were taught, as were his players, to point a finger at the passer after receiving an assist that led to our scoring. It was a lesson in giving credit to the usually unsung, deflecting praise from the overvalued act of scoring. Paradoxically (in the usual Smith fashion), this seemed a fine way to ensure that credit came back your way. Basketball camp, therefore, became a veritable orgy of finger-pointing (the good kind). When the server at the Granville Towers chow line plopped a mound of mashed potatoes on our plates, we pointed at him. We couldn't all be good players, necessarily, but we could all be good finger-pointers.

But I didn't just point my finger. I ran suicides like a G.I. trying to catch a departing helicopter. I screened and cut and passed. I eschewed the dribble. I didn't hog the ball. I knew that Smith valued teamwork above all else, that he sometimes told the story of coaching a phys-ed team at Air Force, where, as Smith put it wryly, he had a player who "shot only when he had the ball." Smith escorted the shooter's four teammates off the court. "Okay, play them by yourself," he told the gunner. "Who takes the ball out of bounds, sir?" he asked Smith. "Good," Smith told him. "Now you've learned that you need two, anyway."

I did have my suspicions that, despite what coaches say about the value of sharing the ball, there was danger at basketball camp in being consigned to the unseen role of hustler, like being a helpful spinster aunt in a Jane Austen novel. You ended up blending so well into the woodwork that you weren't actually seen. Better, perhaps, to shoot a lot, though if you were going to do that, better to hit the shots.

Sad to say, I don't think Dean Smith caught a glimpse of my heroic screen-setting. I did attract the notice of Bill Guthridge, but that was for whistling "Stairway to Heaven" while waiting to be subbed into a game. "Way to whistle," he said. He winked at me, patted me on the shoulder, and kept on walking.

There were times when Dean Smith's story seemed more familiar to me than my own. Though he would never admit it, it was something of a saint's tale—a saint who snuck cigarettes before games in the entryways of gymnasiums, who drank liquor and got divorced and kept a messy desk and loved winning and hated losing, and who kept score with a long memory for jibes and doubters and cheaters. "Dean never forgets anything," the journalist Bill Brill told me, "and he won't let you forget it, either." Brill recalled that when North Carolina won Smith his first national championship in 1982, Dean's first order of business in the victorious press conference was to respond to Frank Barrows, a writer for *The Charlotte Observer*, who had suggested several years before that Smith's teams made such a fetish of consistency that they would never be able to achieve the peak performances necessary to win it all.

And yet, all of these qualities just made his goodness seem more plausible to us everyday sinners. There was something old-fashioned about learning morality through the example of a human being—not from books or sermons or stone tablets, but from an actual life in front of you. Someone who was grinding through the same existence we all were, and yet demonstrated the allure of goodness, who made a case for an ethical behavior rooted in compassion and fellow-feeling—not in the fear of hellfire and damnation, nor in the dictates of the Bible as the ultimate rule book, as was so common in my native South even to this day.

I asked Seymour to what he attributed Smith's decency. He thought for a moment, then said, "I attribute this to a sound upbringing in a good family. He had a family that gave him the good sense to care about others." "Value each human being," Smith's father used to tell him.

Smith extended the realm of family to his players, upon whom he exerted an abiding moral influence as a compassionate but tough-minded paterfamilias. Not even Michael Jordan, whose freedom to do what he wished must have felt near-absolute during his glory days with the Chicago Bulls, who as a sporting demigod enjoyed immunity from simple necessities like waiting in line—not even Jordan could escape Smith's moral suasion. It was as if his former coach had become a voice in his head, a conscience that spoke with that nasally Kansas twang.

As recounted by David Halberstam in his biography, *Playing for Keeps*, Jordan, then a few years into his career as a Bull, arrived late for a preseason exhibition game at Carmichael Auditorium on the UNC campus, where parking was hard to come by on the best of days. Jordan drove into the parking lot next to the gym, only to discover every space filled, except for a single handicapped spot. "Why don't you take it?" his friend Fred Whitfield asked.

"Oh, no, I couldn't do that," Jordan answered. "If Coach Smith ever knew I had parked in a handicapped zone, he'd make me feel terrible. I wouldn't be able to face him."

The question that Smith seemed to answer with his life was how you competed in this dog-eat-dog society, how you competed and even won without turning into a monster. Clearly, Smith hardly minded winning. In fact, he hated to lose.

He revealed that his composure concealed powerful extremes of emotion. "Sometimes I get so upset at myself," he said, speaking in the present tense. "I don't let it show, but I'm really mad." He cited a game that happened nearly 15 years ago against Georgia Tech, in which the Yellow Jackets' Dennis Scott beat UNC at the last second with a three-pointer, falling out of bounds. Smith blamed himself for not having coached a better end of the game.

"Win or lose," he said, "I go to sleep. Then I wake up the next morning and it hurts worse." He seemed relieved not to have to go through that anymore. "I don't read the paper after a loss," he added.

"So how did you learn to lose?" I asked. "It seems coaches would go insane if they didn't."

The answer, Smith suggested, lay in a notion called "the power of helplessness," a concept introduced by the writer Catherine Marshall, whom he'd come across in his thirties when reading religious thinkers like Søren Kierkegaard, Karl Barth, Dietrich Bonhoeffer, and Martin Buber, among the many suggested to him by Bob Seymour. "Crisis brings us face to face with our inadequacy," Marshall had written in a book called *Beyond Our Selves*, "and our inadequacy in turn leads us to the inexhaustible sufficiency of God. This is the power of helplessness, a principle written into the fabric of life."

Smith discovered Marshall's work the same week in 1965 that he was hanged in effigy—twice—in front of Woollen Gym by Carolina students unhappy with the direction of the program under his leadership. A few days after the latter hanging, the Tar Heels upset sixth-ranked Duke at Cameron, 65 to 62. Invited to speak to a celebrating crowd of students outside of Woollen upon the team's return, Smith declined. "I can't," he said. "There's something tight around my neck that keeps me from speaking."

He learned how to apply Marshall's lessons directly to coaching. Giving up in that context meant teaching his players to surrender to the present moment in practice and in games, not to fret about something that was beyond their control in either the future or the past. They were to let go of what they could not control. He helped them—and himself—sidestep the self-immolating demands of victory at all costs. "When we talked to our team over the years," he told me, "our emphasis was always play hard, play smart, and play together. We didn't mention winning. The emphasis was on process versus end result." Many would find it ironic that a coach famous for controlling everything he could ultimately believed in letting go. But the most important lesson of all, one that had generated decades of loyalty from his former players, was Smith's equal treatment of every player, from benchwarmers to stars. He wanted to show them that their value as human beings was separate from their performance on the court.

"The emphasis on process—was that just a sneaky way to get the result you wanted, i.e., winning?" I asked.

"You could say that," Smith said with a Mona Lisa smile. I didn't know whether the smile suggested a realization of the contradiction between not thinking about winning so that you could win, or whether it was simply an acknowledgment that coaches were in business to win and that if they didn't, they'd shortly be in the broadcasting business. But that you might as well try to win in a dignified and decent manner. And lose that way, too.

In this part of the world, coaches were philosopher-kings, their evident wisdom on such matters as full-court pressure and team dynamics eventually translated into realms far afield; both Smith and Mike Krzyzewski had written books applying their coaching secrets to business success. Not long ago, Smith had finished his manual, called *The Carolina Way: Leadership Lessons from a Life in Coaching.* Krzyzewski had published a book entitled *Leading with the Heart: Coach K's Successful Strategies for*

Basketball, Business, and Life. Journalists sought their opinions on matters like the war with Iraq. Politicians asked for their help in fund-raising.

And yet for all of that, one of the things I admired about Dean Smith was that the American vogue for self-promotion appeared lost on him. He didn't court publicity in a country where hype, good or bad, had become a value in its own right. I doubted he had his own blog, for instance. Or that he would be making a commercial for American Express in which he said, "I don't look at myself as a basketball coach. I look at myself as a leader who happens to coach basketball."

He inveighed against the way TV, for instance, zeroed in on the sideline antics of coaches. He had once asked a CBS executive to stop showing coaches during their college basketball broadcasts. "But it's good for the game!" the astounded executive responded.

"Too much is made of this coach-versus-coach business," Smith told me. "It's a player's game."

As for the Duke–Carolina rivalry, his initial position appeared to be: What rivalry? This was vintage Dean Smith. In time, he qualified his view.

He admitted that Duke under Vic Bubas had proved a tough opponent in the Sixties. But in the latter part of that decade, Smith's program caught fire, in part when he out-recruited the Blue Devils for the high school All-American Larry Miller, a six-four bruiser from Catasauqua, Pennsylvania. "He was the big breakthrough," Smith said. "By the time he was ready to sign, Duke had a class of six committed. And Larry said, 'Gosh, I won't get any playing time on the freshman team if I go there.'" From 1967 to 1969, the Tar Heels won three straight ACC Tournaments at a time when only the tournament champion advanced to NCAA play. And for those same three years, UNC went to the Final Four, losing in the championship game in 1968 to UCLA and Lew Alcindor. Duke began a slow fade from sight.

The following decade, Smith said, "There wasn't really any Duke–Carolina rivalry." He thought for a moment. "The team that's beating you is the team that becomes your big rival." In the Seventies, that meant North Carolina State, with David Thompson, had been Carolina's most detested foe. Then Maryland for a time. Then, in the early Eighties, Virginia, with Ralph Sampson at center and Terry Holland as coach. (Holland had named his dog Dean because—as he put it—the dog whined a lot.) "After Sampson left . . . I was trying to remember . . . Duke . . . they didn't beat us much."

And then Smith said, as if trying to recollect something lost in the recesses of either time or his mind: "Duke . . . I suppose they did come on about '85 or '86."

How Smithian, that "I suppose"! In a sense, Smith's reluctance to put Duke into a special category was the cagiest, most rivalrous move of all. By not considering your archrival your archrival, you were injuring him at his most vulnerable point: You refused to give him the prominence he gave you. You were saying he simply wasn't that big a deal! You were saying there was a long list of competitors of which he was one and that perhaps at certain times in the historical record, Duke may have amounted to a pretty good team. But that in the long march, Duke was as often as not standing at the curb, watching the parade go by. Masterful! Ingenious! Devious! And brilliant!

In 1985, Duke successfully out-recruited North Carolina for Danny Ferry. "When we lost Danny Ferry," he told me, "that was the start of their program." This was the same Danny Ferry whose SAT performance was to become a matter of public comment by Smith.

"John Feinstein was furious with me about that," Smith said of the Duke-educated journalist, author of *A Season on the Brink*, a rather terrifying glimpse into the mind of Bobby Knight. "He said it was the worst thing that I commented how Reid's and Williams's SATs combined surpassed Ferry's and Laettner's. I said, 'How do you know that all of them didn't score 1400 or 1500?' Feinstein was so much of a Duke guy."

The controversy occurred the same week as the 1989 ACC Tournament finals, which, in karmically just fashion, ended up being between Duke and North Carolina. The game was a bitterly contested slugfest, bodies bouncing about the lane, every shot challenged. At one point, Krzyzewski stared down the sideline at Smith and yelled, "Fuck you," a remark that was instantly entered in the unexpurgated ledgers of the rivalry. The outcome wasn't decided until Ferry barely missed a long heave at the end of the game. The Tar Heels prevailed, 77 to 74, a win that doubtless gave Smith great satisfaction.

By the time he resigned in 1997, Smith had gone on quite a run against the Blue Devils. "I'd have to check," he said, "but I think Duke beat us once in about ten times before I retired."

"That is exactly right," I said. I knew. Those sorts of things mattered to me, too.

I had come to visit a saint of the everyday to learn about how to be good and how to lose with grace, and had discovered a coach who was still competing, sitting down there in his very own basement giving as good as he ever got—or better. Let the games continue.

THE COLUMNIST AND THE DUD

One morning early in my stay, my mother and I were drinking our morning coffee on our respective couches, both of us steadily working our way through the Raleigh *News & Observer*. My mother always pulled the sports section out first and slung it across the room to me. I cracked it open and that day came across a column by Caulton Tudor featuring the resounding headline: RANDOLPH A DUD AT DUKE.

This shocked me, not because Tudor was wrong (he was undeniably right) but because the *N&O* had been so bold as to pronounce such a controversial, though obvious, truth. After years of living amid the tabloid artillery shells of New York City—headlines and articles that went screaming across the sky and landed at the breakfast table with enough concussive force to rattle the brain along with the silver and the crockery (or, in my case, the plastic and the Styrofoam)—I found the local coverage of sports tame.

I wondered whether that North Carolinian propensity for modesty and sobriety accounted for the relative delicacy of the reporting. Niceness was a supreme virtue here. It was *not* nice to say mean things about people—at least, not directly. This struck me as a situation not especially conducive to truth, an ethics constructed on that inane maxim "If you can't say something nice, then don't say anything at all."

So how strange it was this morning to open up the paper and find an unquestionably negative piece about Duke and Shavlik Randolph. The traditional short paragraphs of the sportswriter marched down the page on their way to an execution. Tudor led off with the bald assertion that "Shavlik Randolph would have been better off at NC State." No hemming and hawing here; Tudor gripped the bully pulpit with steel gloves. He went on: "Then again, Randolph would have been better off at North Carolina. Or Wake Forest. Or Clemson, Miami or South Dakota State."

South Dakota State! Whoa!

Randolph was averaging only 6.4 points and 5.1 rebounds a game for the Blue Devils, then ranked sixth in the country. But just like that home on the range, never was heard a discouraging word from the relentlessly upbeat Randolph. Tudor, however, was another matter.

"Show me someone who insists that he wouldn't change anything about his life, and I'll show you someone living in denial," Tudor wrote. "We all have experiences, decisions, episodes that we regret."

Tudor acknowledged that at the time of his commitment, Randolph's selection of Duke made sense, particularly given the rumblings out of Chapel Hill about Matt Doherty, and the perceived weakness of Herb Sendek at North Carolina State.

But no longer. "The Shav we've seen in college isn't the player we saw at Broughton," Tudor argued. "That player was a graceful wingman with a deadly jump shot, as well as a devastating weak-side rebounder at both ends of the court." Mike Krzyzewski had turned "a potential Larry Bird into a poor man's Larry Lakins," the columnist contended. "It'll be said that Krzyzewski brought in Randolph for no better reason than to torture Sendek on one front and Carolina on the other. . . . Duke hasn't been that great for Randolph, but Randolph hasn't been that great for Duke."

And with a final one-sentence paragraph, Tudor nailed home his point. "It's fair, in that regard, to say that both sides made a mistake."

These weren't exactly fighting words, but in a journalistic culture that usually proved as temperate and undemanding as, say, an ongoing series on a Boy Scout paper drive, Tudor stood out.

Where was I? I wondered. For a moment, I experienced a profound sense of dislocation, like a man waking up in a motel room after 20 nights in 20 different cities. I no longer knew what planet I was on. Tudor's column glowed with a radioactivity common up North, but which had been outlawed as unsafe down here.

To account for this, I went over to Raleigh to meet Caulton Tudor at a Mexican restaurant in Cameron Village. "I'll be wearing a red cap; that's how you'll know me," Tudor told me.

And there he was, the red cap brimming with a boyish head of white hair. Having worked for Raleigh papers since the Fifties, Tudor had been around long enough to become something of an institution, though he carried that reputation as lightly as a five-iron slung over his shoulder. He had already grabbed a booth in the corner. "You're hiding out from the Duke fans," I told him.

"I might need to," he said with a smile. Tudor was at once a jaunty and exquisitely polite man, possessed of a small-town decency, seemingly untaken with himself. His eyes, however, sparkled with what struck me as insurrectionary merriment. His formulations were crisp and bemused. Born in 1948, he'd grown up Angier, North Carolina, a little town 20 miles south of Raleigh. It was then a community of a thousand or so, most of them Wake Forest fans, including Tudor. "Angier had one stoplight, one barbershop, one policeman, one gas station, one school, and 18 Baptist churches," he said. "So that's why we were all Wake Forest fans.

"In those days, that was pretty much typical of eastern North Carolina," Tudor said. "In the small towns, the Baptists were Wake Forest fans, and the Methodists were Duke fans. Neither North Carolina nor North Carolina State had yet gained the large followings they have now.

"If Duke and Wake Forest were playing the following week," Tudor said, "my preacher would always conclude his prayer at the end of the service by saying, 'God bless the Deacons and deliver them from that unholy Methodist officiating.'"

Why, I wondered, had basketball insinuated itself so thoroughly into the North Carolina consciousness? "The dominant culture in North Carolina in the Fifties and even the Sixties was agriculture, and tobacco in particular. People worked like hell in the heat of the summer and in the fall. Then, in the wintertime, they listened to college basketball on the radio and went to basketball games at the high schools. I grew up in that culture. I played high school basketball. I never played a game with the building not full."

Then something momentous happened, according to Tudor. "In 1957, television discovered college basketball—in particular, the three-overtime national championship game between North Carolina and Kansas." The contest had featured the Jayhawks' Wilt Chamberlain against the Tar Heels' feisty New Yorkers. Tudor remembered staying up to watch the contest with his parents. "I was pulling hard for the Heels, even though I thought Wake Forest got robbed in the ACC tournament when they played North Carolina."

The success of the local teams, stocked mainly with imported players, created legions of new fans and swelled them with pride. "So that's how everybody in this area was bred," Tudor said. "Even today in this region, with its multicultural conglomerate of people, if Duke or Wake Forest or North Carolina are on television, the kids will be in the house, looking at a game. That hasn't changed. It's still the common denominator."

The columnist dated the ferocity of the Duke–Carolina rivalry quite precisely: to Mike Krzyzewski's arrival at Duke, in 1980. "It wasn't really there before Krzyzewski," he said. "Krzyzewski made that rivalry. It was his quest to bring down the king."

The king, of course, was Dean Smith. And in Tudor's opinion, the 1984 ACC Tournament game between Duke and Carolina, when Duke managed to eke out a 77-to-75 victory after losing both regular-season games, stood as perhaps the third-most-important game in ACC history, after North Carolina's 1957 championship contest and the 1974 game between North Carolina State and Maryland ("the greatest game I've ever seen," Tudor said). "I don't think Krzyzewski deep down thought at that time he could catch North Carolina. If he did, he thought it was going to take a long, long time. And I don't think Duke fans at the time had an ounce of confidence in Krzyzewski.

"I'm convinced that Michael Jordan shot North Carolina out of it," Tudor said of the 1984 game. "He just wouldn't quit shooting. Sam

Perkins kept holding his hand in the air, calling for the ball. Jay Bilas was trying to guard Perkins, and no way that was going to work. That was the day that everything turned, I think."

Irony of ironies, now Krzyzewski was becoming the very man he had striven to bring down all those years. "Obviously, they're both great coaches," Tudor said. "But what's funny is that the older Krzyzewski gets, the more he is turning into Dean Smith. Krzyzewski had a great line: He said that he never wanted to get to the point where he couldn't go to the corner 7-Eleven in his blue jeans and buy a carton of milk and a six-pack of beer. Well, now he's there. He's turned into Krzyzewski Incorporated like Dean turned into Dean Smith Incorporated."

As unpleasant as the coaches might find the media's intense interest in them, they ought to thank their lucky stars for it, Tudor said. "If we didn't cover basketball like the Second Coming, these coaches would be making the same thing as the women's softball coach. They certainly wouldn't have a private tower. They'd have a little office like the track coach."

It wasn't just the coaches who benefited from the fervor over basketball. "There's no way to tell how much money Krzyzewski is worth to Duke. There's no way to measure. He himself is probably making three million a year when you add everything up—shoe companies, personal appearances, advertisements, camps, salary. But beyond that, there's no telling what he's brought that school in corporate endowment. If you look, Duke has a couple of franchises. It's got an educational franchise, but not like Harvard and maybe not like Stanford. But it also has a basketball franchise. If Harvard lost its law school, for instance, it would erode the value of the university. It wouldn't happen overnight, but it would happen."

Tudor found the Shav column difficult to write. "I knew at the time it would create a lot of discussion," he said. He turned his palms upward and shrugged, as if to say, What could I do? He said the piece had provoked at least two hundred e-mails and so many phone calls that his voice-mail system had been overloaded. Dozens of handwritten letters had also arrived, with more trickling in. Joe Alleva, the Duke athletic director, fired off two or three e-mails to Tudor, accusing the columnist of being unfair to the program and unduly hard on Randolph.

Lots of readers were irked, he said, but others had agreed with his appraisal. "I strongly suspect that a lot of the Duke fans were just upset that the program and Krzyzewski had been criticized. Duke fans are very defensive about their image."

Shavlik Randolph had been the most highly recruited North Carolinian ever to have chosen Duke. At the time, his choice struck me as a potential sign of an epochal change in the way young Carolinians viewed themselves. He had chosen personal advancement over home. Under Dean Smith, it hadn't been necessary to make such choices; a player from North Carolina could have both in Chapel Hill. But Randolph's decision for Duke meant that either Duke had become part of North Carolina in a way it never had been before, or that North Carolinians no longer felt the same loyalty to their state institutions. Their notion of home had become more elastic. In retrospect, Randolph's commitment in 2001 probably represented the high-water mark for Duke within the state and nationally. And perhaps Randolph's failure to succeed there meant that Duke would once again be no better than second choice for talented local ballers. The pipeline that had been briefly opened had once again been cut.

THE SCREAM
UNC versus Maryland
The Dean Dome
January 8, 2005
Noon

The first home conference game of the season for North Carolina featured a great matchup at point guard between Raymond Felton and Maryland's John Gilchrist, both equally bullish in the upper body. If either player managed to dip a shoulder ahead of his opponent on the way to the basket, the defender might as well have resigned. They looked as if they were engaged in a body-to-body game of nok-hockey.

North Carolina had been Gilchrist's favorite school, but the Tar Heels had recruited Felton instead, and now Gilchrist was playing as if he intended to show UNC it had made a mistake. He put a saucy move on Felton and hit a jumper. As the half progressed, his game started to seem less old-school than reform school; he wasn't playing North Carolina, he was playing Raymond Felton. The other players had disappeared, the court had shrunk to cracked asphalt playground, and Gilchrist's game had turned into *I can top whatever you do*. Roy Williams had bled some of the *mano a mano* out of Felton's attack, sometimes to the team's detriment, but at times like this you could see the virtue of Williams's teaching. To put it in the words of my friend Doug Marlette, who observed ACC basketball through a keen psychoanalytic lens, the rampaging, out-of-control id (Gilchrist) was being contained and rebutted by the team-oriented superego (Felton).

Here's what it looked like: On one end, Felton passed to May for the dunk. UNC led, 16 to 12. On the other end, Gilchrist nailed a great bad shot over Felton. Stalking the sidelines in his typically frantic fashion, Maryland coach Gary Williams was sampling Edvard Munch's *The Scream*. He knew that the bad shot that went in now was inducement for Gilchrist to take the bad shot that missed later. His point guard roamed the court, an outlaw within shouting distance. All Gary could do was scream and sweat on the sideline. And roast inside his suit like a clay-baked piece of fish.

The bad shots were coming. Even with the score at 19 to 18, UNC's favor, even after Melvin Scott had canned a three and Maryland had answered with a two, Gary Williams could *see* them coming. He could sense them even when Maryland took the lead by one on a monster dunk with 12:20 left in the first half. The Tar Heels had not been down even a point for longer than I could remember.

And Maryland even extended its lead, going up by seven, Gilchrist hitting a difficult two along the way. Then, with 8:27 to go in the half, he was called for a foul on a charge. Felton then popped in a long three— he was on a tear from the outside—and you could see the blood in Gilchrist's eye. And suddenly, UNC went on the sort of run it had been going on all year, punishing opponents with speed and shooting. Pick your poison. Rashad McCants stole the ball and alley-ooped to Felton for two. Marvin Williams hit a three. McCants called for noise, as was his wont, while heading back down the court, and the roof nearly lifted off the Dean Dome. By halftime, North Carolina was up 47 to 34. Gilchrist headed to the locker room like a beaten man.

"It's already over, Gilchrist! Might as well not even come out for the second half," the beast screamed.

"Give the poor kid a break," the journalist said. He was also concerned about whether anyone might have witnessed this breach of journalistic decorum.

The second half quickly turned into a decimation. The Heels shredded the Maryland press, constantly reaching the basket with two-on-one and three-on-two advantages. UNC was making Maryland their bitch—and Maryland wasn't that bad a team! Inside, Sean May wore Maryland's Will Bowers behind him like a backpack. No matter which way Bowers moved, May had him perfectly sealed off, unable to get around to his front or side. When May hit two free throws with 7:45 left in the game, he gave the Tar Heels a 30-point lead at 88 to 58.

John Gilchrist had been buried on the bench for much of the second half. A rumor ran through the crowd that Gary Williams was refusing to talk to his players. Melvin Scott scored on a layup with contact to make it 107 to 65, then Reyshawn Terry amplified the victory with an alley-oop. The final score: North Carolina 109, Maryland 75.

The way Maryland had been spun around in this game seemed aptly expressed in the second half when forward Nik Caner-Medley was greeted at the free-throw line by the chant, "Shorts on backwards [clap-clap-clapclapclap], shorts on backwards [clap-clap-clapclapclap]." Even the beast felt bad for the poor fellow. But not that bad.

Together, the beast and the journalist headed for the locker room. The beast wanted to see the sweat-plastered mummy of Gary Williams, and the journalist agreed that this would be a worthwhile sight. On the front row in the media room in his usual spot, *The Chapel Hill News*'s Eddy Landreth was expounding on what happens when a team played at the pace we had just beheld when Gary Williams entered. His face was flushed. He seemed drained of life by the vampiric bite of the game.

After finishing the postmortem on his team and himself, Williams called over to the longtime Tar Heel radio play-by-play man, Woody Durham.

"Woody, if you see John Bunting [the North Carolina football coach], tell him I'm sorry I didn't shake his hand. I think he was coming up to me after the game trying to shake my hand, and well, ah, I missed shaking it." Early in their coaching careers, Williams and Bunting had been colleagues at Rowan University, a Division III school in New Jersey. But Williams was in no mood to shake anyone's hand after that pounding.

Williams extended his hand to show how Bunting had tried to greet him. He studied the hand as if it belonged to someone else. Yes, that must be my hand, he seemed to conclude. Williams appeared to be coming back from a long trip out of his body. It was clear that it took a while for the intensity of the game to leave him. Once it did, however, he appeared to be a kindhearted guy. I knew all about this sort of thing.

"Okay, good buddy," Durham said, with the compassion of a fellow who worked in close proximity to men whose lives were literally defined by winning and losing.

By contrast, Roy Williams bustled into the media room, displaying the ruddy glow of a winner—an instant golf tan. He was "extremely impressed with the team," he said, "with the exception of a ten-minute stretch in the first half that resulted in a broken clipboard."

"I was disappointed with Rashad and Melvin at that time," he said. "I got all over them, but I loved the way they came back in the second half."

In the players' lounge, reporters surrounded Melvin Scott, jamming mikes and tape recorders at him as if probing for radioactivity. The Carolina senior had scored a respectable nine points, nailing two threes along the way. But the main reason the press formed a semicircle around Melvin was because he was always ready with an interesting quote. And there was good context this afternoon. Maryland was in his backyard. He had spurned Gary Williams for North Carolina, and this win made it a lot easier for him to go back to Baltimore. "This was for bragging rights," he said. "I remember our losses to Maryland." At College Park his freshman year, Maryland had practically skunked the Tar Heels, 112 to 79. "You go through all that for a reason. This silences the critics."

I wasn't sure who these critics were. But it was always good for an athlete to invent them and then shut them up.

"If we took care of the ball better, the lead would have been even bigger,"

Melvin said, enjoying himself, his postgame pizza cooling beside him. The guys never ate in front of the reporters. "You was like, *Wow,* when you saw the score, right? I'm not trying to brag but we're even better than that."

"Melvin, how about when Roy broke his clipboard?" asked one journalist.

"He was just writing real hard and it broke," Melvin said, teasing the press. They laughed. His earrings sparkled. They seemed to reflect a finely balanced glee.

If he was still adjusting to coming off the bench, it didn't show this afternoon. "Me, Marvin, David, and QT, we call ourselves the Four Horsemen," he said of the prime reserves. "Maybe we'll think of including Reyshawn if he keeps coming like today."

Outside in the hallway, Melvin's mother, Bridget, stood waiting for her son. She was wearing a blue warm-up suit and a white basketball jersey that said "M Scott #1" on the front and "Mom #1 Fan" on the back. "To be honest, in my heart, Melvin coulda got twenty," she told me. "They looked him off a whole lot." She meant the other players, especially Felton and the big men inside.

I made it back to my mother's in time to see JJ Redick actually miss a free throw in Duke's game against Temple. He knocked in the next shot. Duke was winning 82 to 74. That turned out to be the final. They should have beaten Temple by more. This struck me as a very good sign.

THE BALTIMORE BLUES

One winter weekend, I went to visit Melvin and his family at home in Baltimore. No place could have been more different from Chapel Hill. Baltimore was a nineteenth-century town of twenty-first-century pain. Half-tumbled buildings of bygone grandeur, bricked-up and plywooded windows, crummy trees, teenage drug tycoons, junkies condemned to wander the streets at all hours like ghosts of their former selves—the place seemed to encourage both languorous heroin nod-outs and frantic amphetamine hustles. Horizons tended to lie close at hand: a few blocks, a set of stoops, a humble room. Baltimore looked like the perfect stage set for TV dramas of urban decay: drugs, hustle, and murder. But it was real. People lived and died in Baltimore a lot faster than they lived and died in most places.

Melvin and I had first met one afternoon at the Smith Center. He had come to lift weights and was carrying with him a Styrofoam container with two chili dogs, one of which he offered me. In context, his graciousness struck me as exceptional. Most players at schools like North Carolina and Duke had already spent close to a quarter of their young lives being pursued by the many supplicants of the basketball kingdom: coaches, fans, shoe companies, summer camps. So heavily were they courted that it was hard for even the most natively generous not to regard themselves as the human equivalent of Polynesian statues ringed with offerings to the gods.

Melvin, too, had been highly sought, his prowess as a high school player trumpeted by the likes of the respected recruiting writer Frank Burlison: "There may not be a better jump shooter in the national class of 2001," he wrote on FOXsports.com. And yet, Melvin seemed not to take his good fortune for granted. It may have been a measure of how far he'd traveled in his journey from Baltimore and Chapel Hill, a five-hour car trip but light-years from Melvin's neighborhood.

The Scotts rented a three-story house made of gray stone at 1827 North Calvert, just a few feet off North Avenue, which divided the city into an East and West Side. For many years, they had stayed "up the hill" on the other side of North Avenue, at 20th and Barclay. It was a tough neighborhood in a tough town, and it was there that Bridget Scott had raised two daughters and four sons. She also had another daughter that she'd allowed to be raised elsewhere.

For 23 years, Bridget Scott, now 51, had been a junkie and an alcoholic. She drank, she smoked, and she injected heroin. As she put it, she was out there in the world. But she always took her children to church, where she sang in the choir while floating on a sweet dope cloud, looking out at Melvin and his brothers and sisters as the congregation joined in the singing. Then on March 2, 1988, she'd had enough. She went into a detox program, and she'd been clean ever since.

Now she worked as a janitor at a public school, mopping floors and emptying trash. Her children adored her. When the boys were still playing high school basketball, they often left with her after their games and helped her clean until nine or ten at night.

Melvin and his mother showed me the letters from college coaches still taped to the wall of his old bedroom. "I got even more in drawers," he said. His mother pulled pictures of Melvin, graduation shots, an image of Melvin lying in the middle of the court at the Dean Dome while still a high school student. "He always wanted to go to the school where Michael Jordan went," she said.

In Baltimore, Melvin toured me through the streets of the neighborhood in which he'd grown up, past the old court at Public School 39, past the street where a fellow baller he admired had gotten into an argument about a game and had been shot. When the man crawled under a car to avoid being shot again, he'd been shot several more times anyway and had died. "Yo, brother Melvin," yelled a man riding by that spot on a bike. "You got to put it up more." Everywhere we went, people kept asking Melvin why he wasn't shooting more, why he wasn't driving to the hoop the way he used to do. *What's going on, man? You got to pick it up, brother!*

In his senior season at North Carolina, Melvin found himself poised uncomfortably between his old life and his new, between his former dreams of basketball stardom and his current position on the bench. As we walked, he wondered aloud whether life in Chapel Hill was so sweet that it had robbed him of his motivation, which came from the streets of Baltimore, from his desire to rebuke the doubters that had dogged him every step of the way. He was too small, they said. Too wild. Too poor. Too disadvantaged. He felt the need to represent the doubters while simultaneously showing them up. The irony was that in Baltimore now, they thought he was soft, a Chapel Hill guy. And in Chapel Hill, they thought he was street, a Baltimore guy. So who was he?

That evening, Melvin, his friend Isaiah "Zeke" Johnson, his brother Will Scott, and Melvin's nephew, Tay, walked around the corner to Kentucky Fried Chicken. The wind sliced across North Avenue, and everybody hunkered down into their jackets. Swallowed by their hoods, they looked like a procession of inner-city monks on a fast-food pilgrimage.

In the KFC, a skinny, tipsy, middle-aged lady they had passed on the way asked the fellows for change. "I gave you a buck and a dime," Melvin told her.

"I need more to get a soda and a box of chicken," he said.

"Soda comes with the box," Melvin said.

"Does a Crown Royal come with it?" the lady said.

Everybody laughed.

They walked back into the cold. The snow crunched underfoot. Melvin was carrying the bucket of chicken. He spotted a Cadillac stopped at the intersection of North and Calvert. "That's that old-school nigger," he said. The window rolled down and Melvin shook hands with the old-school nigger.

The girls laid the table with the fried chicken, slaw, mashed potatoes, and biscuits. Everyone held hands and Bridget prayed. "Thank you, Lord, for bringing this family together." Melvin and Zeke heaped their plates with food and went into the living room. One of Melvin's nephews was running around with a toy helicopter making loud whirring sounds.

Like two codgers, Melvin and Zeke, both in their early twenties, began reminiscing about the old days. Zeke came from Up the Hill, Melvin from Down the Hill, but they'd liked each other from the get-go. They'd been running together for 18 years, raising pigeons, squeegeeing cars on North Avenue, and eventually hooping together. At Greenspring Middle School, the two boys had played two years together, with Zeke being a year ahead of Melvin. Their second year, the team went 25-0, winning the city championship when Zeke hit a buzzer-beater after rebounding Melvin's missed jumper.

Now 24, Zeke was tall and rangy, a forward completing his last year of eligibility at Bowie State. He'd already played at four schools: New Mexico, Palm Beach Junior College, Charleston, and now Bowie State. As he said, he'd been loyal in his life "to three things: my kids, basketball, and the street." He'd hustled, gambled, robbed, and sold drugs. He had needed money because he wanted little Zeke to be well-dressed in a way he himself had never been. He'd fathered two children when he was still in middle school. "If you're from, like, the suburbs, you might flip about that," he said, "but it's not weird around here. Two of my brothers had kids when they were thirteen." He was an enthusiastic poet of his experience, tugging on a wisp of a goatee as he told tales full of nostalgia and rue. You'd have thought him twice his age. He'd already packed several lifetimes into one. And this cold winter night, at the ripe age of 24, he was looking back in mystification at the ways of the young.

"The kids these days are different," Zeke said. "When we were coming up, we made our own fun. We made things out of *nothing*. We didn't have all these video games and cell phones."

Between sips of a Mountain Dew, Melvin nodded assent. "That's true," he said. "These kids don't have a clue as to what tough times is. We found ways to entertain ourselves. We were survivors. Now a kid's cell phone gets cut off and he thinks it's the end of the world. My mama would have smiled. 'You got a job?' she would have said. 'No? *Then you ain't got a cell phone!*'"

"You have to have heart to grow up in Baltimore," Zeke said. "Remember that time in middle school I got robbed of my shoes on the way to school? $120 pair of Timberlands. My mother bought me those shoes."

"I saw this nigger here walking through the school with no shoes," Melvin said, laughing, "and I said to myself, 'What happened?'"

Zeke said: "The principal and the staff members were like, 'Why you playing, Zeke? Why you got no shoes? Why you go and be playing with me, Mr. Johnson?' I was so mad, I went off on everybody."

Melvin was laughing. Zeke was still outraged. "I said, 'They took a gun to me and took my boots!'"

The guys who robbed Zeke rode the bus every day with him and Melvin. They were slightly older, jealous that the high school girls would talk to the younger guys. "We were cool. We'd be on the bus, rocking. We used to sing and rap," Melvin said. "*Give me the loot, give me the loot!*" On the Carolina team, Melvin owned a reputation for being the best freestyler, capable of improvising a rap at a moment's notice.

"The girls liked us," Melvin said. So much so that one of the guys accompanying the girls said to Melvin, "Yo, I'm just saying, don't talk to my girl no more."

"I ain't talking to your girl," Melvin said. But he knew what was coming next, a beat-down. The same guys had pounded one of their classmates until thick pools of blood had been left drying on the ground. He and Zeke started taking three different bus routes to get to school to avoid the older guys. Sometimes they didn't make it to class until one o'clock in the afternoon. Sometimes they didn't go at all until after school was over, and they went then just to hoop, as Melvin put it.

"That's when basketball was basketball," Melvin said. "Not all these politics like now." Everything changed for Zeke and Melvin once they hit their respective senior years of high school. It would never be the same after that. A sport they had played for love of the game became the means to an end. They became servants of basketball rather than the reverse, moving from the freedom of first romance to the forced march of a marriage. "Back in middle school and the first years of high school, it was just love—love of the game," Melvin said. "I still love basketball, but back then I used to *really* get into it. It was crazy!"

"Just hooping," Zeke said with a sigh. "It was before you really knew what basketball was about. We had teachers involved in middle school. They'd be in the gym holding signs with our names on them." He still regarded this as astonishing, all the love for the players. His mother didn't come to his games. "But Miss Bridget came like I was one of her own children," he said.

In the summer, the fellows would start playing at noon and keep going until two or three in the morning at different courts around the city, Greenspring or 39. Midnight Madness, they called the late games. And afterward, they'd hang out next to the blacktop, talking, laughing, chillin'. "You didn't worry about nothing else," Melvin said.

At Greenspring, a gangster everybody was afraid to foul would sometimes show up for a quick game. "That's all right," he'd tell the boys. "Play me hard. Don't worry about it."

One time, Melvin and Zeke, overconfident, lost a game of two on two to Ty and Billy, a couple of older competitors. The price of defeat: They had to take off their clothes and run home through the streets of Baltimore, naked.

"We had to come down from 24th to 20th," Melvin said, laughing. "They told us we could keep our undershirts on but not our drawers."

"I walked past my mother like that!" Zeke said, his voice rising with the indignity. "And these girls were chasing us. Those guys that won said, 'We're not playing. If we catch you, we're gonna knock your blocks off.'"

"We were young and dumb," Melvin said.

"I thought we were going to beat those guys," Zeke said. "I thought that was when we were coming into our time."

"You weren't even thinking of college at first," Zeke said. "Then you get that first letter. That's big." His arrived from Syracuse, Melvin's from Wake Forest and Dave Odom, the head coach then.

"But my guy was Tommy Amaker," Melvin said of the coach then at Seton Hall. "Tommy's cool. He told me the truth. If he was faking me out, he was good. He said, 'I see you play, and I see these other guys, and you're the one doing help-side defense.'" Melvin flushed with pride at the memory. "And I'm like, 'That's what Smitty, my high school coach, taught me.' Tommy was the best recruiter ever, better even than Matt Doherty."

Amaker had fallen in love with Melvin's game when he was head coach at Seton Hall, but he left for Michigan. Enter Matt Doherty, who was no slouch at recruiting himself. He watched Melvin play a night or two after he got the North Carolina job in a game between D.C. and Baltimore, featuring high school stars such as Carmelo Anthony and James White, a player North Carolina had been pursuing.

It was an audition, Zeke said, a cobbled-together event just to get all that talent in one gym. Gary Williams was there from Maryland, Steve Wojciechowski from Duke. The coaches were excited, he said, prodding and provoking the players. "They were screaming, 'Go at him!'" The Baltimore players set ball screens for Melvin. He scored 30 points, handed out 11 assists. It didn't matter that he missed the last shot of the game, or that Baltimore lost. He'd made an impression. Matt Doherty shook his hand. "James White was nowhere to be found," Zeke said, ever loyal to his running buddy Melvin.

In his last year of high school, Melvin took a recruiting trip to North Carolina, flying by private jet to Raleigh-Durham. He wore a suit, his hair done up in cornrows. On Saturday, he played pickup with Joe Forte, the Tar Heel All-American, among others. "My ball," Forte kept saying, calling fouls on Melvin. "I didn't foul you," Melvin said. Forte wouldn't leave the court until he won. Then Melvin attended the football game at Kenan Stadium, worrying that he stank because he had left his deodorant behind in Baltimore. Students held up signs with his name on them, begged him to come to Carolina. "Oh, my goodness," he said, still pleased at the memory. "I don't know how they knew who I was." That night Melvin went out with some of his future teammates to a nightclub, Players, where he took a drink—more than a few, actually—for the first time in his life. "I was dancing away. I'm like, 'I'm a man!'"

Melvin cracked up, remembering.

That weekend, Matt Doherty had asked him, "Are you going to make me the happiest coach in America?"

Melvin said, "I'll let you know at the end of the weekend."

"I won't be able to sleep," Matt Doherty said.

Zeke laughed at the dialogue. "That's corny, man," he said. "They some salesmen, those coaches." He imitated Doherty talking to a recruit. "*Me and my wife were talking about you. I won't be able to sleep until you commit to the program.*"

Melvin's brother Charles barreled into the living room, with a "Melvin Scott #1" jersey on and dreads trailing from his cap like waves of energy. He had just returned from playing at the Y and was still emitting competitive electricity. "I took a charge to win the game," he announced. With his arrival, the atmosphere in the living room instantly transformed from languid to crackling. Everyone who knew Charles agreed that he was lionhearted (Charles said it himself), that he hated to lose, that he fought until his last breath. Everyone said that if you took Charles's heart and Melvin's game—his shot and fundamentals—you'd have an unstoppable player.

Charles had preceded Melvin at Southern High School by three years, and they had played one season together in the same backcourt, Charles at the point, Melvin at shooting guard. After high school, Charles won a scholarship to Connors State in Warner, Oklahoma, but he hadn't felt appreciated out there in that unimaginable place so far from Baltimore. So he quit. He came home, and before he could get his life squared away, he was arrested for murder. A guy had hit their sister, Rozetta, and Charles had talked to him. A week later, the guy turned up dead. Charles stood trial twice for the murder, both times getting off on a hung jury. He was innocent, he said. "The first time I won 8 to 4," he said. "The second time it was 7 to 5." Now he was always telling Melvin not to follow his example. Never quit, he told him. The worst mistake of his life was quitting Connors State. He wanted Melvin to play with more heart. He loved his brother with an almost motherly devotion. If Melvin made it to the pros, he planned to follow him and help him any way he could.

But sometimes, Charles had a funny way of showing his love. "Can anyone check Melvin Scott?" Charles asked the room. And then he answered his own question. "You ain't got to check him. He check himself."

"*No*," Melvin said in a rare show of resistance. "*You* can't check him."

"You ain't got good enough sense to go to the hole and jump stop," Charles said. He pounded his chest. "This is where it starts. Heart. One thing about heart, it'll take you a long way."

Melvin sank down into the chair, silent. He put his hand over his face in a gesture that said he was tired. Not this again.

Zeke came to Melvin's aid. He started busting on Charles, referring to an old game. "You kept saying to me, 'You got Melvin.' Why you not check him?"

Charles ignored this. "You know who has a big heart?" he said. "Julius Hodge. To me, he ain't that good. But he got a lot of heart."

"He even talk when he ain't playing good," Zeke said.

"On the court, ain't nobody your friend," Charles said. "Off the court, different story."

"How you get heart?" Zeke asked. He was as street-savvy as they come, tough when he had to be. But he displayed a sweet openness rare for Baltimore guys, who tended to shield their doubts behind a stoic façade. "I lost the swagger," Zeke admitted. "I had the swagger in high school. I had the fire then. But in college, they don't like you to have the swagger. I got on the bus one time with my sweatsuit on when we were supposed to have our suits on. Suspension, dawg. You ain't got to do nothing really wild in college to get in trouble."

Swagger was everything for a Baltimore guy. Swagger took you through bad neighborhoods, poor games, and seasons of uncertainty. Swagger was a way of launching into the unknown, attempting acts of heroism you didn't know whether you could complete. Swagger was style, swagger was fortitude, swagger was a way of walking the knife-edge of hardship. Not like a man who owned this mean old world. More like a man who didn't give a damn.

So when his swagger went, a young guy's life changed. A floor collapsed from under his heart. The horizon closed in fast. He started to get old, no matter how old he was.

At Zeke's moment of doubt, Charles said something wonderful, something that put the swagger back inside Zeke, that in one fell swoop exalted the real reason Zeke should have the swagger, which had nothing to do with triumph on the court, Charles's usual standard for success in all things. "Zeke, I'm proud of you," he said. "You could have given up anytime." He reached across the room to slap Zeke's hand.

Zeke beamed. "That's a nice present, Charlie," he said. "That makes me feel good. I feel so good, I might have to finish school!"

Then Charles started back in on Melvin, who by now was protecting himself by appearing half asleep, slouched in the chair, his eyes heavy-lidded, his head leaning to one side, resting on his hand. His jeans hung down, revealing that, like Michael Jordan, he wore UNC shorts under his pants. His earring sparkled. On his wrist he wore a bracelet that said "Jesus."

"He got skills," Charles said, referring to his brother. "But he ain't got heart. My heart bigger than his. My first step quicker than his, too."

Now even Zeke was joining the dissection of Melvin's game. "The athleticism in his legs is not the same as when he was in high school," he said. Everybody in the room, Bridget included, could anatomize a player's strengths and weaknesses, break them down like a scholar.

"It's a mind thing," Charles said for the umpteenth time.

Without changing his position, so that his utterance had the effect of a dead man's suddenly speaking, Melvin said, "Basketball gonna stop one day."

"I want to finish my degree," Zeke said. "Get a job and be a family man. Maybe play foreign basketball if it pays the same as a job."

"What's your plan, Melvin Scott?" Bridget asked.

"Well, I'm taking school seriously," he said. "Because one day ball is really gonna stop. Maybe get bread overseas if I can."

"I wouldn't want you to come back to Baltimore," Bridget said. "People will try to tear you down."

"I want to help people learn," Melvin said.

"So many people envy you fellows," Bridget said. "They wish they would have did the work to be where y'all at."

"People want to know what I'm doing now," Zeke said. "I'm embarrassed to even talk to them. They like, 'What happened?'"

"But them niggers doubted you out a long time ago," Melvin said.

"You guys hung in there," Bridget said, "through all those trials and tribulations. People counted all y'all out a long time ago."

The room was quiet, everybody gauging the distance they had come, the distance still to go, the adjustments they might have to make in their expectation of what life held. Basketball had been king for so long. As always, Bridget was the voice of endurance, of persistence and gratitude. Referring to another of Melvin's brothers, she remarked, "Kevin gets upset because people around here say, 'What's wrong with Melvin?' But this is just negativity." In Bridget's lexicon, "negativity" was a word on par with "hell" and "heroin" and "giving up," words she understood intimately in all of their dark allure but finally rejected.

"What are *they* doing?" Bridget asked. "Nothing but staying on the corner. I tell Kevin, 'We got to realize this—what are *they* doing? The negative ones, the ones that say that stuff about Melvin—what's going on with *them*?'"

Everyone in the room was listening intently to Bridget. She preached to a congregation eager to have their doubts allayed, and she'd be damned if she couldn't give them what they needed. "We come from 20th and Barclay!" she exclaimed. (Like saying you come from the corner of Hell and Damnation.) "Melvin is the first young man from Baltimore City to attend the University of North Carolina! What does that *mean*?"

Zeke caught the spirit. "My brother came from nothing," he said. "The same nothing I came from. Now he's setting an example for all the brothers, and he's doing real good. He makes fourteen dollars an hour. When I see a guy taking care of his kids and his home, I see I ain't got to be rich! Guys say, 'You got to make it to the League.' But you don't got to do that."

"A true friend doesn't care if you don't make it to the League," Bridget said.

Melvin said: "I went out one time and watched all those guys surround Melo when he came back to Baltimore. What's crazy is those niggers didn't think we were gonna be nothing."

"Yeah," Charles said, as lionhearted as ever. "But ain't nothing like getting paid for what you want. I'm gonna be in the League. Watch."

Nobody told him he wouldn't. Nobody said he would. Nobody had anything to say. Nobody wanted to kill such a fragile thing as hope. It was a crime to take the swagger out of a man.

It was getting on towards midnight. The cold had settled over Baltimore like a lake. Rousing himself from the chair, Melvin walked outside, escorting one of the evening's guests to his car, parked down Calvert next to vacant lot. "Charles is right, man," he said. "I need to go harder. I wish he would be with me in Chapel Hill all the time. Chapel Hill is too easy. I lose my edge there. It's not the city; it's not Baltimore. It's easier to have an edge up here."

His freshman year, Melvin had started 21 out of 27 games. Last season, he made 27 starts in 30 games. And now, not only did he begin games on the bench, he didn't even play that much. But he refused to knuckle under.

"I'm not giving up," he said. "Just stay in there, fighting." He turned and trudged down the street through the snow, following in his own footsteps, and went back into the house.

FOUR

Must Have Been the Dress Shoes

THE TRAITOR

BY JANUARY, the saga of Shavlik Randolph that had begun with a bang heard round the state seemed to be ending in a midseason fizzle of recrimination and bewilderment. Randolph's signing with Duke in the fall of 2001 out of Broughton High School in Raleigh had appeared to represent both an epochal moment in the Duke–North Carolina rivalry, and, in retrospect, the high point of Duke's recent run of domination. That's because Shav was a North Carolina boy, and good North Carolina boys—highly recruited ones, anyway—did not go to Duke. In fact, in the nearly 20 years since David Thompson of Shelby's Crest High School spurned Dean Smith and the Tar Heels for Norm Sloan and the Wolf-pack in 1971, the state's top players, if academically qualified, had rarely ended up anywhere but Chapel Hill.

And then along came Shav, as he was known not just to friends and family but to the entire basketball universe and just about every conscious citizen of North Carolina. Like Madonna, like Jesus, all Shav needed was one name. Anything more was redundant. Watching him as a high school player, scouts touted him as the next Larry Bird: He shot well from the outside, could slide inside for layups and rebounds, passed like the Indiana trickster, and—oh, yeah—he was white. Others compared Randolph to Tim Duncan for the quiet but deadly efficiency of his game.

And yet, with nearly half of his junior season gone, Shavlik Randolph was averaging career numbers of seven points and four rebounds. Had he been overrated? Undercoached? What happened? I had a crank theory that I wanted to try out.

One afternoon, I drove over to Duke to visit the big man. It would be fair to say that I hadn't wept very much over Shav's troubles. Perhaps I had wept not at all. The beast within me knew that such woes were ultimately good for North Carolina, that whether or not it was reasonable, his failure to meet expectations would redound not only to his own but to Krzyzewski's discredit as well. I was looking at the long-term, shall we say, *evolutionary* consequences of such matters. Given Shav's sterling reputation coming out of high school and his lack of success at Duke, future recruits, particularly tall ones, might believe their games would develop better at some other university—a school very much like North Carolina, for instance.

On the other hand, I felt considerable sympathy for Randolph. His senior year in high school, a profile by G. D. Gearino had appeared in the Raleigh *News & Observer*, in which it was revealed that (for at least one memorable evening), the 18-year-old Randolph had let his mother cut his steak into pieces for him. A nicely grilled steak it may have been, but it was raw meat for the partisans of North Carolina, who were already sulking that the local star had picked Duke. The news of Mrs. Randolph's knife work spread across the Internet faster than the "I Love You" virus.

On the drive over, I recalled a conversation I'd shared with Kenny Randolph, Shav's father. The first thing Kenny Randolph had told me about his son was that "his walk with the Lord has kept his head on straight." He added that Shav was "like an angel to our family. People love and respect him."

Kenny was garrulous and proud, delighted to portray his son in a mythological light. He described his recruitment with relish, relying on a fine-grained memory for nearly every drop-step, put-back, and turn-around. He called the AAU summer circuit the best time of his—*Kenny's*—life. "Playing games out in the country, deep in South Carolina," he said with a sigh. "Just for the love of it. The kids were twelve or thirteen at the time. Then, a couple of years later, everybody knew who they were!"

They especially knew who Shavlik Randolph was. Even given their habit of slathering hype over just about every prospect who could tie his own shoes, recruiting analysts reached daunting levels of hyperbole in describing him. "Simply put, he's the most skilled big man in the class of 2002 and he plays like it almost every time-out," wrote Dave Telep.

"Watching Randolph play the game of basketball is quite simply a thing of beauty," wrote Clint Jackson. "Okay, I've seen enough. This boy can play!" rhapsodized another scout after witnessing Shav elevate a foot and a half over the rim (or so legend had it) and posterize a hapless opponent with a dunk.

The speculation on his college choice began early, but Kenny Randolph tried to stay away from the topic for as long as possible. "His whole junior year, I never mentioned a thing about where he should go," Kenny told me. "But the week before his senior year, I said, 'Son, this is out of control. Narrow it to five schools.'"

So Shavlik trimmed his list to the Triangle schools—North Carolina, Duke, and NC State—along with two out-of-state universities. One was Kansas, then coached by Roy Williams, who obviously still had strong links to the area. The other was Florida, whose coach, Billy Donovan, had shown up one morning at Broughton and stood silent (coaches weren't allowed to speak to recruits during that period) in the high school parking lot so Randolph could see him. (I recalled Caulton Tudor saying of Donovan, "He scares me." And it was true: The image of Donovan standing in a high school parking lot, mute, staring at Randolph, induced shivers—it was *The Morning of the Living Dead*.)

Once Shav had narrowed the list to the five finalists, Kenny told him, "I think you need to forget about moving out of the area."

"What are you talking about?" Shav said.

"Why would you want to leave this area?" his father asked. "Duke and UNC, the two greatest programs in history, are here in your backyard. If you went to NC State, it would be like a fairy tale. You were born and bred in Raleigh, and you will come home to Raleigh after basketball. You're a homebody, son. You're just that kind of person."

So, in the fall of 2001, Randolph went on his official visit to Duke, the school widely viewed as the favorite for his services. "He got home from the visit on Sunday night," Kenny said. "I was in bed, watching a ball game. Shav lay down beside me and said, 'Dad, I got to talk to you. I know where I want to go to school.' Like I was surprised."

Shav told his father, "Something's telling me I need to go to Duke."

The news of Randolph's decision flashed across the state like a bulletin from the Emergency Broadcast Service. The unwritten headline: North Carolina homeboy picks Duke! Realignment of the stars! From Murphy to Manteo, the reverberations were immense. Families headed to their cellars. A strange light glowed on the horizon. An unnamed comet approached. Fish fell from the sky, the waters boiled. Portents of a dark age! Duke had supplanted North Carolina as the destination of choice for the state's best player. For a place with its identity so wrapped up in basketball, home would never be the same.

Given all that, when Shavlik Randolph arrived at Duke in the fall of 2002, he was lugging a gym bag weighed down with expectations. His first game for Duke, against Army, in which he scored 23 points and collected seven rebounds, suggested that he could carry that load. So did the next game, against Davidson, where he rang up 17 points and 12 rebounds.

His third game, against UCLA, began to tell a different story. He played only eight minutes, earning no points and a mere two rebounds. For the season, he ended up averaging seven points and just under four rebounds a game. The bottom fell out and had been falling ever since.

It turned out that Shav had been suffering from a bad hip, which had caused problems for his left foot. He lost his explosiveness. The Duke trainer had tried to address his pain with orthotic supports in his shoe, but Shav preferred his own orthopedics, a cut-up tennis ball. After his freshman year, he'd undergone surgery to alleviate the problem. But by then, Duke's lack of big players meant Krzyzewski wanted Shav to bang inside. So he bulked up in preparation and lost his quickness, which was marginal to begin with. Too often, he found himself caught flat-footed, a step slow on defense and mired in perpetual foul trouble as a result. And if that wasn't bad enough, in December, he'd been diagnosed with mononucleosis and missed four games.

Now, midway through his junior year, he was averaging 5.9 points and 4.7 rebounds. This was not what anyone had expected, least of all Randolph himself. "On Tobacco Road," Kenny had said to me, "people put these athletes on a pedestal." Shav had been winched and craned onto a really big pedestal. Did he jump off his pedestal, or was he pushed? Either way, it was a long ways down.

The Shavlik Randolph I met that afternoon at Cameron impressed me as a sweet-souled, gawky young man. His hair stuck up from his head in little sprigs, less wild than boyish, the kind of thing a mother would pat down with her saliva. Whatever pain he might have suffered from his physical ailments seemed matched by the piercing realization that the world had turned on him in a strange, inimical way. His relation with his own talent was uncertain now. He was the caretaker of a deteriorating mansion that had once been the envy of the neighborhood.

Friendly enough, he also appeared a tad wary, submerged in a world of thought, a tad more reflective than athletes at their leonine prime, for whom thinking is simply unnecessary, if not a downright obstacle.

Before I got around to raising my theory with him, we spoke about his life at Duke. Currently, he was learning how to talk to girls. His tutor was his freshman-year roommate, Sean Dockery, with whom he sometimes double-dated. They made an intriguing duo—Dockery, black, a six-two point guard from inner-city Chicago, and Randolph, white, an angular six-nine Southerner from Raleigh, North Carolina. In exchange for Dockery's schooling Shav in the finer points of male-female interaction, Shav instructed Dock, as he was known, in the niceties of Christianity.

"When I first came to Duke, I had been so focused on basketball," Randolph said, a bit abashed. "The other wasn't really a part of my life." He had a hard time actually saying the word "girls." "Sean and I are really different, coming from two different backgrounds, but we really bonded," he said.

Watching Dockery operate had been instructive. "He's a big ladies' man. He knows how to talk to girls. He kind of showed me the ropes. But every now and then, I would be like, 'Dock, you shouldn't do that. That's not something God wants us to do.'"

"Has he given you any specific advice?" I asked.

"He says just to be confident," Randolph said. "You've got to know who you are. And no one's more confident than he is."

Randolph shook his head and smiled at Dockery's brass. "He'll walk up and just start talking to some girl," he said. "He's not afraid of rejection. I think that's what separates him from the pack. And I've never seen him be rejected."

Given Randolph's awe at Dockery's gumption, I suspected that his own psyche was spun out of much more fragile glass. I wondered how easeful it was to play for Mike Krzyzewski, who could fill a player's ears with burning caustic. I had heard that Shav didn't like that Coach K cursed so much. In the war between Mike Krzyzewski and the rest of the world, I wondered if Randolph was too compassionate a soldier. I began to dance around the topic and soon discovered that in this waltz, anyway, Shav was a nimble dancer himself. Either that, or he was telling the truth.

"You seem to have a very different personality from Coach K," I said.

"I think that's fair to say," Randolph said.

"Is his intensity ever a problem?"

"It's one of the main reasons Coach K has been so successful," he answered, pirouetting around the query. "He has all that intensity and will to win inside of him, and he's able to make his players feel like that. He develops a personal relationship with you, and therefore, you want to fight for him."

"What sorts of things does he say to you?"

"He says I should go out there and not worry about making mistakes. He says he knows I can do this. Just let go and play. Every possession, try to work. Think about rebounding, rebounding, and rebounding. And playing good defense. The offense will come."

The word floating around the Triangle basketball community—journalists, hangers-on, friends of the players—was that Randolph and his father were unhappy that Shav had been forced by the Duke staff to slug it out under the basket. He and Kenny saw the forward as a wingman who wasn't allowed to showcase his gifts. "Do you miss playing on the perimeter?" I asked. I described how in a 2004 game, I'd watched him take Wake Forest's Eric Williams off the dribble from the perimeter and swoop in for a rousing slam. That wasn't the sort of thing you saw much from Randolph, then or now.

A trace of competitive ire seemed to flash across his face. "The team hasn't really needed me on the perimeter to shoot threes," he said.

I asked him if he was still happy with his choice of Duke. He said that he was, that he liked being part of a winning program, that he liked the student life at Duke.

Another rumor drifting between Durham and Chapel Hill had it that Krzyzewski and the Duke assistants were unhappy with Randolph's lack of aggression. They thought he played soft. This led me to the topic that I felt might explain Randolph's curious lack of success. It was an odd theory, yet it seemed plausible. "Do you feel that Christianity might affect your play?" I asked him. Many players thanked Jesus profusely for their talents, especially after a good game, but Randolph appeared to practice a more profound Christianity, in which one had obligations to realms outside of one's own talent. Christianity was not just a developmental league for aspiring professionals.

It struck me that Christianity teaches you to be compassionate, loving, to put other people first. On the court, however, you sometimes had to be selfish, aggressive—in a sense, un-Christian. At the very least, you had to put yourself before your opponent. In basketball, the last won't be first; the last will be cut. I asked Randolph, "Couldn't Christianity get in the way of a player's ambition to be great on the court, to beat the other guy, to win?"

"Yes," Randolph said simply, somewhat to my surprise. "That's definitely something that any Christian athlete is presented with. It's definitely an issue for me."

I hadn't even mentioned to Randolph that hatred might come in handy. It seemed to me that players such as Michael Jordan hated intensely—for at least the duration of a game—whoever got in their way. They hated failure. They hated losing. They hated anything that summoned up vulnerability or the times they'd been losers.

He thought for a moment. Then he qualified his views. "For me," he said, "I wouldn't be where I am today if it weren't for putting God first in my life. And I would drop basketball in a second if that's what I felt God intended for me to do. But I feel like He intends for me to play basketball. I go out there trying to do His will in mine, and try to do what I have to do out on the basketball court to be successful without doing anything immoral. I represent God when I play. So if I have a good performance, I try not to take credit because it wasn't really me that did it."

This didn't sound like boilerplate coming from Randolph. He was as sincere as his coach was profane. There were times when it appeared as if he'd be happier as an Amish farmer, ordering all aspects of his life through his faith, withdrawing from this world, leaving all the corruption and compromise behind. Still, basketball was what he did, what he was good at.

And, he added wryly, "Even if you're playing for God, that doesn't mean you can't be competitive. When David slew Goliath, I'm sure he threw the stones pretty hard."

This was a good point. Melvin, too, identified with David slaying Goliath. I envisioned a whole team of college players called The Davids, all intent on proving themselves to the doubters. Not theological doubters—no, the critics who thought they'd already reached their high-water marks in basketball. For Randolph, his current status as a bruising role player presented a complicated religious question. It must be a test from God, since everything in his life came from God. Not just his success, which he credited to Christianity. God had brought him along as a player. God would see him through.

"Things didn't really start happening for me in basketball until I lifted it up, and that's when it started happening," he said.

"When you accepted Christ?"

"Yes."

I asked him how old he was when this happened.

"I was fifteen and a half," he said.

At Duke, with classes, practices, and many games on Sunday, Randolph didn't have time for church, but he prayed every day. "I've ended up putting my trust in God," Randolph said. "Not to say that once you become a Christian, you're going to be successful at everything. I've definitely had my ups and downs. But no matter what, God's been there for me. He's helped me do things that I wouldn't have been able to do without Him."

There was still one more thing I wanted to know that afternoon. And it had nothing to do with God. "About your mother cutting your meat," I ventured.

Before I had finished my question, he grimaced and ducked his head, then looked up at me with an expression of exhaustion. "Yeah, some people are like, 'Shav, does your mother cut your meat?' What they don't understand is that in practice that day, I jammed three of my fingers. She had cut a couple of pieces of my steak because I couldn't do it. I don't know if you've ever had a couple of jammed fingers . . ."

"Oh, definitely," I said.

"It makes it hard to cut your meat. The photographer there that night who took the pictures for the newspaper—every time I see her, she apologizes."

I wondered if there was any place else in the United States where at that moment a journalist (or beast) was engaged in clarifying a mother's motivation in cutting her son's steak in an attempt to achieve a glimpse into that son's success or failure as a basketball player. They would understand this in Kentucky, I figured, and maybe Indiana, two states similarly obsessed with college basketball.

"So you feel the whole thing got blown out of proportion?" I said.

"There's worse things that could happen than people saying my mom cuts my steak," Shav said, quite reasonably. Or at least it would have been quite reasonable had he lived anywhere else on the planet.

PRACTICE
The Dean Dome
January 13, 2005
4:00 P.M.

Driving to the Dean Dome, I found myself stuck behind a jogger who was running down the middle of the road at the bend near Forest Theater. He never turned back to look at the long line of cars forming behind him. He never detoured over to the sidewalk. This struck me as suggestive of an overly idealistic worldview that deserved to be complicated by a good tap on that professorial ass with two or so tons of Subaru.

I arrived an hour into practice. Things seemed relaxed, easy, as befitting last night's win against Georgia Tech, 91 to 69.

In his white practice jersey, Melvin deflected a pass that had been attempted into the middle of the lane. "Goaltender, thatta boy," Williams yelled at Melvin. The coach was patient and professorial this afternoon.

Roy was talking about last night's game. "We had twenty offensive rebounds, and we didn't capitalize on them," he said. "If we can learn to do that, we'll be hard to beat."

Now the Tar Heels were practicing late-game situations. One minute left to play, 20 seconds left on the shot clock. Raymond Felton scored on a drive after Sean May set a screen.

Melvin came in and took Felton's place. "Same scenario," barked Roy. "You're up two with one minute to play. Got to be something really good."

Jackie Manuel missed a long three. That was not something really good. "At least you took twenty seconds off the clock," Williams said.

Williams had promised that they'd run one more set, and David Noel nailed a three. "I lied to you," Williams said. "I want one more."

On the next sequence, Jawad Williams scored inside. "Jawad, that's what I want," the coach yelled. "Like last night. Use the rim so that Luke Schenscher can't get to it."

As usual, practice ended with the players running 22's. After Reyshawn Terry made a false start, Williams said: "I'm going to let you out, Reyshawn, because for the first time in the history of Carolina basketball, a senior has volunteered to run. Melvin! Melvin Scott!"

I think it is more accurate to say that Melvin was volunteered, but no matter: He ran by himself on behalf of the team. He ran hard and effortlessly, and in the last 20 feet, he turned and ran backwards, highstepping his way home. Roy Williams laughed. Melvin screamed. And practice was over.

As I left practice, rain was pelting the parking lot at the Dean Dome. It was still coming down as Duke took on NC State in Raleigh a few hours later.

WHEN THE TIME IS WRONG
UNC versus Wake Forest
Joel Coliseum
Winston-Salem
January 15, 2005

It was the best of times, it was the worst of times, it was the season of light, it was the season of dark, but more than anything else it was the season of "male erectile difficulty." Cialis had taken over the airwaves during sports events; commercial time appeared a competition between beer and pills. Viewers were invited either to deflate that portion of the male anatomy subject to deflation or winch it back up for when the time was right. The Cialis commercials came with a titillating warning: *In the rare event an erection lasts more than four hours, please consult your doctor.* While I suspected that this was a case of priapism as loss leader, four-hour erections weren't the kind of high-rising action I wanted to hear about while I viewed games with my mother and my sister.

As it turned out, the actual game this afternoon was even more disturbing than the commercials. Over the past month and a half, I had become increasingly convinced that I might have returned to North Carolina just in time to witness at close range one of the greatest Tar Heel teams ever. UNC featured a rich blend of athleticism, skill, experience, and depth. It was clear that the freshman forward Marvin Williams was a major talent. At six-nine, he rebounded well but was also capable of drifting outside for three-pointers. His willingness to come off the bench meant that chemistry was not a problem for a team returning all five of its starters. After the surprise loss to Santa Clara that opened the season, North Carolina had won 14 straight games, including victories over two ranked ACC teams, Maryland and Georgia Tech, blasting the competition into the stratosphere. They were averaging 93.7 points a game, best in the country.

Sometimes they toyed with their opponents, as if asking them, Do you want your beating now? Or would you prefer to wait a few minutes?

This afternoon, the third-ranked Tar Heels hit fourth-ranked Wake Forest in the mouth, and the Demon Deacons popped them right back. This was not in the script. In fact, Wake kept popping—three-pointers. Taron Downey sank three in a row, giving the Deacons a first-half lead that they never relinquished. He ran down the court, jawing at Rashad McCants, who played only ten minutes in the half due to foul trouble. For the first time all year, North Carolina was down at the break, 43 to 33.

In the second half, Wake Forest continued playing terrific defense, sagging in on Sean May at the post, daring Carolina to shoot from the perimeter. McCants, UNC's best outside threat, stayed in foul trouble, and while he would finish with 19 points, he missed huge chunks of the game. Melvin, who had led the Heels to victory with his shooting at Wake the year before, was playing only sporadically. Late in the game, the Deacons surged to a 17-point lead. Time for a commercial.

"If a moonlit moment turns into the right moment," the Cialis voiceover intoned. My mother, my sister, and I stared in different directions. "There are some good recipes in this *Hearty Soup Book*," my sister said, breaking the silence.

Gloom and vexation pervaded the room, despite my sister's attempts to spark a discussion of soup.

When the game resumed, the Tar Heels made a final charge. With 4:51 left, they had whittled the margin to 13. McCants drove to the basket, converted a layup, and got hammered by Wake's Justin Gray. It appeared that he would go to the line for the chance to add a free throw. But the ref waved off the basket and directed McCants to the line for two shots. It was that sort of afternoon. Carolina cut the lead to seven, but that was as close as the team got. With the Wake Point point guard Chris Paul playing brilliantly, squirting through cracks in the defense, scoring and delivering the ball to his teammates, there would be no comeback. North Carolina's David Noel, Jackie Manuel, and Marvin Williams all fouled out. The Deacons took advantage of the closely called game, nailing an ACC record 32 of 32 free throws. The final score was 95 to 82.

My mother flicked off the TV. "That was dreadful," she said. "They've got some work to do."

"It was bad," I said. We all sit quietly for a few minutes.

"So, Annie," I finally said. "Let's hear one of those hearty soup recipes. And it better be a good one."

E.T. MALONE'S DREAM

"Is this Will Blythe?" the man on the other end of the phone asked. "The guy who hates Duke?"

"That would be me," I said. It hadn't been my life's ambition to be known as the guy who hated Duke. But that's how things seemed to be working out. People may admire a person who's compassionate and kind, but they love a good hater.

"My name is Ted Malone," the man told me. "I'm an Episcopal priest. And I hate Duke, too." He sounded like an addict declaring himself at a 12-step meeting. It seemed that I was becoming a sort of clearinghouse for fellow obsessives.

Hello, my name is Ted M., and I am a Duke hater.

Hello, Ted!

He told me he heard me on WCHL, the local radio station, talking with D. G. Martin about my obsession. He'd had a dream, which he wanted to tell me about in greater detail, and which had led him to draw a picture that had now been manufactured as a poster you could buy at Sutton's Drug Store on Franklin Street. He thought I might be interested in all of this because the dream and the poster related to Duke and North Carolina.

I was interested; frankly, he had me at "Hello. I am an Episcopal priest and I hate Duke, too." So I got directions and that afternoon drove several miles south of town. Malone's compact house was nestled into the woods on a gravel drive off the highway between Chapel Hill and Pittsboro. Crews were widening the road, and great clouds of dust hung in the air as if the landscape were being consumed by invisible fire. I felt I had entered the outskirts of a biblical parable of exodus. The development struck me as the kind of hidden place that grad students and young couples passed through on their way up and out and never again revisited.

Malone met me at the door. He looked to be in his sixties. Broken-up sunlight dappled the living room, which seemed to be the province of a quiet man who liked to read and who collected things. Through the woods, we could hear the bulldozers and the pavers and the dump trucks grinding away. Malone's residence appeared to be the cottage of a cultured Southern bachelor, but over the course of the afternoon, he kept fondly mentioning his wife, Lynda, whom he had met when she was only 12 and he 16, at Wrightsville Beach, where his family spent summers in an old wooden apartment building they owned. "My brother said she was still in diapers," he said with a chuckle.

He offered me a glass of port. "I'm having a glass," he told me. "Would you care to join me?"

I did. He poured liberally. And he and I sat in his living room, surrounded by the old things that meant so much to him, and he began to tell me about his dream and his life. Malone had a beautiful, soft Carolina accent. His words curled around your brain like a sleeping cat. The effect was to induce calm. You found yourself getting very sleepy from time to time.

Of course, Ted and I were also drinking port in the bright middle of a weekday afternoon, and that, too, could have accounted for some of the drowsiness. He told me that he'd grown up the only Episcopalian in Harnett County. He'd served in the Army back in the Sixties, bounced around as a reporter ("Every paper I worked for went broke"), taught journalism, and eventually joined up with the Episcopal Diocese in Raleigh "as one of the scribes and Pharisees." In 1991, he was himself ordained as an Episcopalian minister. He loved North Carolina writers and history, and had published a few books of poetry with small presses. He and Lynda had two children, a daughter, Anna, and a son, Ned, the sophomore at Carolina.

Tall and skilled, Ned had been a promising basketball player early in high school. "I thought I was going to be a basketball daddy," Malone told me. Instead, Ned got interested in theater and singing. His senior year, he played on a rec-league team with a group of improvisational actors. The coach wore a cow suit. The players would leapfrog on the court. They lost every game. "I wish he'd told me he was going to be a crack-cocaine dealer," Malone said in that genteel Carolina voice. "But he does have a beautiful baritone."

Recently, he and Ned had watched *The Return of the King*, the third film in the Tolkien trilogy. Malone loved the films and the Tolkien books from which they were made, and the imagery of those works lingered in his brain. So one February night, he went to sleep, and before morning, in a state halfway between asleep and awake, he experienced one last dream, and it was a doozy. In that dream, Malone was inhabiting Tolkien's universe—he was inside *The Lord of the Rings*. But this realm was inhabited not by hobbits, but by Carolina and Duke basketball players. "In my vision," Malone said, "I saw Raymond Felton holding the sword rather than Frodo. And Felton said, 'My sword turns light blue! Dorks are nearby!'"

"'Dorks! Dorks!'" he exclaimed, delighted with the workings of his unconscious. "And instead of Gandalf from *Lord of the Rings*, the wizard was Old Roydalf. And when Felton says, 'Dorks are nearby,' Old Roydalf says, 'Yes. About seven miles away!'"

Here Malone paused to let the connections sink in. "Seven miles away!" he said, a psychic detective unraveling the mysterious processes of his own mind. "That's just about the distance between Carolina and Duke!" The quiet of his little house rang with the unspoken question: *Can you believe it?* Malone was growing excited.

"Then I woke up," he said. "And I thought, *I can do something with that*. I can portray the great battle between the forces of good and the forces of evil. In *The Lord of the Rings*, the Dark Lord lives in the Land of Mordor, and I thought, Hmmmm . . . the *Dork* Lord lives in *MorDurham*. See?"

I saw.

He decided to portray his dream in a picture. He was no stranger to such work; he'd drawn literary and historical maps of North Carolina, Georgia, and Florida that were sold by McGraw-Hill and displayed in English classrooms. ("Spanish moss is a devil to draw," he said. "It looks just like spaghetti.") But those maps, with their authors and book titles, had been mere warm-ups for the cosmological ultimate: the eternal battle between good and evil known as the Duke–North Carolina rivalry.

For two straight weeks in February, he'd stayed up every night until three o'clock, drawing and coloring his vision. "I'd just started a new job that very week at St. Timothy's Church in Raleigh, as a priest there and an assistant to the rector." So he gulped NoDoz and drew. And now he wanted to show me the results.

We went into the kitchen, carrying our glasses of port with us. On the table, Malone unrolled the poster slowly, as if he were opening a Dead Sea scroll in a temperature-controlled vault. He could have been wearing archivist's gloves. And then, at last, there it was: one of those medieval topographies of heaven and hell, yet featuring basketball teams in place of saints and sinners. Malone had divided the cosmos into halves. On the left side, he'd drawn and colored an enormous sun, upon which he'd emblazoned the poster's title, "Return of the Roy." (Scholars, please note the play on *roi*, the French word for "king.")

Under that auspicious sun gathered the Carolina players and their coach. Old Roydalf's hair fell below his shoulders. He wore wizard's robes and carried a clipboard chalked with X's and O's. From his neck dangled a whistle that appeared to have talismanic significance. Arrayed around Old Roydalf were his players. Melvin Scott, here nicknamed Treyo, for what I trust are obvious reasons, guarded the lower corner of the poster. He had been portrayed as a gnome, with terrifyingly hairy feet. "To tell the truth," Malone confessed, "Melvin is off a little bit." Raymond Felton's feet also sprouted impressive bristle. Whether these hairy feet carried iconographic significance in Tolkien, I hadn't a clue.

Malone had given Felton the appellation of Assisto. Jackie Manuel, shown crouching with bow and arrow, was called Defenso. With one hand gripping a broadaxe, the other a catapult, Sean May had earned the tag Boardso. Impish even here, Rashad McCants was Scoro; Jawad Williams, Dunko; and David Noel, lurking behind the Old Well and hoisting a spear, had been designated Leapo.

No one could accuse the artist of obscurity in his naming. On the Carolina side of the cosmos, "Assisto" Felton sounded the alarm in a cartoon bubble: "My sword turns light blue! Dorks are nearby!"

And indeed, there they lurked, the Blue Devils. Or, rather they were retreating in panic to the Land of MorDurham, a dark, sulfurous kingdom (was that tobacco smoke spewing from the volcano next to the Duke Chapel?) commanded by the Dork Lord, Mike Krzyzewski. Malone had rendered the Duke coach as a campy Dracula, though the camp aspect seemed possibly inadvertent. "We hates them Tar Heels, my Precious!" Krzyzewski snarled. His players (no nicknames, only numbers) cowered nearby. Shelden Williams quavered at the approach of Rameses, the Carolina mascot. JJ Redick begged for his "mommy."

In the poster, the rout was on, though apparently unnoticed by the tiny Cameron Crazy who was busily studying a text entitled *Rude for Life* up in one of the campus's pseudo-Gothic embattlements. The Duke cheerleaders appeared to be muscular guys with tattoos. They certainly filled out their uniforms.

It would not have been too much of a stretch to say that the Duke–Carolina rivalry had found its Hieronymous Bosch. We sat in the sunlight of that green afternoon, sipping our port and contemplating the spectacle. Here heaven (North Carolina) and hell (Duke) collided.

"I don't have a lot of reasons for hating Duke," he said, surveying the carnage. "Just a few simple ones. What always comes to mind is arrogance. Some of the people I've known who went to Duke personify a rudeness and abruptness that you associate with Yankees. When my son, Ned, was a senior at Chapel Hill High School, you could have predicted the kids who ended up at Duke. They're the ones who wouldn't help the little old lady cross the street. They'd run her down!"

Yankee. On a fine afternoon almost 150 years after the Civil War, Malone had used the word as a kind of epithet, the way an ever-dwindling number of North Carolinians did. My father had sometimes employed it in the same fashion, as a kind of shorthand not so much for a Northerner as for the kind of American who wanted to get ahead at all costs, the sort who'd—well—run down an old lady. In my experience living and working among a great many of them in New York, Yankees never thought of themselves as Yankees. They might identify themselves as any of a thousand varieties of American, as Italian or Lebanese or Greek or German, as Puerto Rican or Dominican or Sudanese or Nigerian. As Jews or Catholics or Muslims. But never as Yankees. Southerners, however, could spot a Yankee, especially Southerners of a certain age.

Yankees lacked lilt. They barreled around as if life were a list of errands that had to be checked off by five o'clock. The chambers of their hearts were as steely as bank vaults, but empty inside. Yankees measured the world with money. They stared impolitely at things. They got to the point too quickly. They had literal minds. For Yankees, defeat did not constitute a lifestyle. They didn't laze around on the porch. They didn't drink port in the middle of a weekday afternoon and tell long stories that wound around their subject like wisteria vines.

And where did these Yankees go when they came to North Carolina? To Duke, Malone said. "Duke is like a foreign implant. The University of New Jersey at Durham."

There was a heat coming off his rhetoric that might have been out of proportion to its target, though that was always hard to judge when it came to Duke. Still, I knew myself that the bonfires of Duke hatred were sometimes fueled by secret disappointments and rage that had nothing to do with that school. Duke fans were similarly served by their hatred of Carolina, Republicans of Democrats, Democrats of Republicans. And so on. Build a big fire and let it burn with all of the injustice and regret and sadness that ever consumed you.

Then Malone grew wistful. "I'd rather catch a Carolina game with Lynda than anyone else," he said. "Because emotionally and physically, she's completely absorbed in the game. She'll say, 'Come on, Raymond, honey!' She likes Jackie Manuel. He's a wonderful player, by the way, who seems like he needs mothering. He's the one who got Roy's program the quickest."

"Is your wife at work?" I asked. The house was devoid of a woman's touches. He spoke so elegiacally of his wife, with such love and tenderness that it was hard to believe she was actually around. He seemed drowning in nostalgia, actually, and not a generalized longing for a departed past, for the North Carolina of his youth, but for a particular person who was missing. I had the feeling that I had located Malone's secret sorrow. I was abashed to ask, but like a Yankee, I did. I had to know.

He cleared his throat, his face mottling with blood. "Well, actually, my wife and I are divorced," he said. "Since 1996. After 25 years. One reason my wife and I got divorced was that our house was too small. I'm not a pack rat, but I am a collector."

I hated to think that Malone had been ejected into the wilderness (with all of his books and pictures and keepsakes) merely for his propensity to collect. But if there was more to it than that, I had no wish to know. "I still wash my clothes at her house," he told me. "We go to church together sometimes. I still love her. She's the love of my life. I'd like to marry her again."

We sat quietly for a moment. "No matter what happened," Malone said, "Lynda and I, we still hated Duke."

"You'll always have that in common," I said.

This was not the time to tell Ted Malone that the Russian army once executed soldiers suffering from nostalgia. He was too far gone. Like my father, I knew what it was to lose an afternoon to longing. We all suffered from it, that old nostalgia. Men got it worse than women, but then men were usually the ones that screwed up.

We sat a while longer. The engines out on the highway had gone silent. I had learned to let sorrow keep quiet, to pool inside me and still. I set my port glass on the table. Malone walked me to the door. I thanked him for his hospitality. His eyes were brimming with tears. "Keep the faith," I said.

"Go Heels," he said.

MUST HAVE BEEN THE DRESS SHOES
Practice
The Dean Dome
January 20, 2005
4:00 P.M.

The Tar Heels were scrimmaging full-court. Quentin Thomas ran into a pick and, as if he'd been picked off by a sniper, went down hard just past the midcourt line and rolled around on the floor in agony. But Roy Williams wasn't feeling a lot of sympathy for his fledgling and inconsistent freshman point guard. "You get hit by a frickin' screen and lay on the floor like that?" he yelled at Thomas. "We can't do that against Duke or North Carolina State. Get a little tougher!"

Thomas hauled himself off the floor in sections, like a man picking up his own body parts. Grimacing, he limped to the sideline, replaced in the game by Melvin, who proceeded to throw a cross-court pass that was intercepted and returned for a layup by Reyshawn Terry. Williams shook his head in disgust.

Turnovers continued to be a problem this afternoon, as they had been in last night's game at Clemson, when the Tar Heels had thrown the ball away 23 times. Felton sent a pass soaring over McCants's head and yelled, "My bad, Rashad." But Roy Williams wasn't having any of it, contrition or not. "We'll run fifty-five for every turnover," he barked. A 55 was a practice sprint consisting of ten lengths of the court run within 55 seconds.

Williams stalked the court like a banty professor of basketball who was about to keep his class in at recess. The point guards weren't reading the defense correctly. "Everyone!" the coach yelled. "Q, Raymond, Melvin, Wes—if you're playing point guard, you can't just call a play. You've got to see what the defense is doing."

Two possessions later, the starters again lost the ball on a break, firing a pass across the baseline before anybody had arrived to receive it. "Want me to put it in Saturday's game that every turnover is a fifty-five?" Roy asked, exasperated. "Would you like that?"

On the next play, Felton nailed an open three, as he'd been doing a lot this season, and at last, Professor Williams was placated. "Can someone tell me what we did on that play?" he demanded. Rather than waiting for this Socratic dialogue to be consummated, he answered. "We didn't try to make a great pass. We just made good ones, and we got a wide-open shot."

Now it was time to run, and from the sounds of it, there was going to be a lot of running. Roy Williams was notorious for making players run more than even Matt Doherty had done, and Doherty's suicide drills had often left the players vomiting in trash cans strategically arrayed around the court. The team lined up on the baseline and crouched in readiness, waiting for the whistle. At that moment, Williams turned to the sideline and said to an unseen presence, "Michael, remember when you were a freshman, how you volunteered for everything? Marvin and Quentin have volunteered to run for the team today."

And at that moment, striding onto the court as if out of a Nike commercial, came Michael Jordan. Supposedly, he had stopped by Chapel Hill on his way to a golf lesson in Pinehurst. He watched approvingly as Marvin and Quentin ran for the sake of the team. (It may safely be said that being watched by the greatest basketball player in the history of the world appeared to quicken their pace.) When they were done, Jordan called for a ball from a manager. In his khakis, black shirt, and dress shoes, he strutted to the opposite end of the court and began to shoot threes. All eyes were on him. The first attempt clanged off the side of the rim. The second was an air ball. "Oh, oh, oh," he exclaimed, a sheepish smile breaking across his face. The smile, at least, was still the same. Magnetic, playful, visible for miles. And so iconic, even from a distance, that it nearly absolved him of his horrendous shooting. The managers kept feeding him the ball. They were now playing a small part in a historic moment: the day Michael Jordan missed every shot.

Jordan turned to someone on the sideline—it might have been Dean Smith, who had also shown up for his former player's royal visit—and said, "How much you want to bet?" He was still a betting man. He then swooped in for a dunk, khaki pants flaring like wings. And missed it badly. The ball caught on the front of the rim just like it does when a junior high schooler first tries to jam.

Must have been the shoes, I thought. Those shoes just weren't made for dunking.

Jordan turned to the sideline again and this time said, "Ten in a row." He began shooting jumpers. He sank the first, missed the second. And the third. And the fourth. Every time he missed, he growled, "Aahhhhh-wwwww." He hadn't lost any of his fire, that much was apparent. Jumper, yes; fire, no. Finally, he gave up on the jumpers and sailed in for a dunk, arm extended straight up (he needed every inch of extra height now), ball cradled in his massive palm, his leather shoes now a dark, shiny blur as he levitated. I think that I can speak for everyone in the

building (including Michael) when I say that we desperately wanted
him to jam this shot. We craved the orgasmic rush that we had come to
expect when he threw the ball down as if punishing it for insubordina-
tion. We needed Michael to be Michael. Just do it. Please.

And, thankfully, Jordan did it. He put it down. Barely. When he
returned to earth, which admittedly did not take as long as during his
heyday, he was holding his stomach as if he had pulled a muscle. "That's
my exercise for the day," he said. He slipped his watch back on, big and
sparkling even from a distance.

Jordan went over and joined the players where they were doing their
post-practice stretching on mats. They surrounded him like ten-year-
olds at a basketball camp. He was the hero of their youth, the figure at
the center of their childhood fantasies of future domination. They had
been Michael Jordan as they fired jumpers on playgrounds and at rec
centers. Melvin remained in a full squat, raptly attentive.

Jordan spoke with the natural command of a guy who'd been granting
audiences for almost a quarter of a century. I was reminded of something that happened when I had interviewed him nearly ten years
before. He was sitting in a wingback chair at one end of a suite at the St.
Regis Hotel in New York, discussing the variety of scents (pine was one)
that had been blended to produce a Michael Jordan fragrance, a bottle of
which was even now fermenting in my medicine cabinet, because to this
day I simply couldn't bear to discard this promotional graft, this *relic*
that had come to me directly from the great man's hands. "Isn't that
refreshing?" he had asked me that day at the St. Regis as I sniffed the
cologne. "You can really smell that pine," I said.

The signature moment in the interview, however, occurred when Jordan pulled a cigar from his pocket. In that instant, the television personality Ahmad Rashad, who appeared to have no other livelihood than
interviewing Jordan at the halftime of NBA games, swooped to his side
with a little clipper to prepare the cigar for smoking. He took the cigar,
snipped an end off, and receded again to a nearby wall. All of this happened without Jordan's saying a thing. He had merely fished the cigar
out of his pocket, raised it regally toward his lips, and Rashad had

arrived with the clipper dojiggie faster than a player on help-side defense. That is what it was to be twentieth-century royalty.

This afternoon, the great man was telling the guys how he and the rest of the former players lived vicariously through the current team. He confessed that he didn't make it to as many games as he should, but that he watched them all on TV. He said that when the current team lost to Duke or North Carolina State or any team, he and the other former Tar Heels suffered from the losses. He jibed the players for the many turnovers they had committed during the first half of the Clemson game. It always felt good to come home and visit with Coach Smith and the rest of the staff, he said. He assured the fellows that Carolina would always stay with them no matter what they did in the future. He wished the team well in its quest to finish this season as national champions.

Jordan hadn't always been so enthusiastic about the core contingent of this year's team. Two seasons before, when Matt Doherty had resigned under pressure, Jordan fumed that "there's no way that eighteen- and nineteen-year-olds should be dictating the future of a coach." He groused about "new-jack players."

Today he finished his pep talk to a round of applause. He then threw a basketball at Rashad McCants's chest. McCants beamed. He and Jordan exchanged mock elbows. To have been teased by Michael Jordan, to have had the ball fired into your chest at close range by him, to have taken a playful elbow from the man himself, that was about as close as a young player could come to being anointed.

FIVE

The Dog Days of Winter

THE EDUCATION OF A LONER

"OH, WAITER," MELVIN SCOTT CALLED OUT in the dining room of the Carolina Inn. "Would you please ask that young man to join us?"

That young man, seated by himself in a far corner of the room, was Rashad McCants, star swingman and All-American enigma. McCants had an answer for his teammates. "I don't want nothing to do with those fools," he told the waiter. And he stayed where he was.

The occasion was a team get-together during then–Carolina recruit JamesOn Curry's official visit to the campus in the fall of 2003, but the exchange was typical. McCants was a loner, often at odds with the world and perhaps even with himself. Before the current season, he'd gotten two tattoos etched into his biceps. One read "Born to Be Hated." The other, "Dying to Be Loved."

But as solitary as he could be off the court, he was just as singular a presence on it. More than any other Tar Heel, including the point guard Raymond Felton, McCants combined consummate inside and outside games. At six-four and a robust 207 pounds, he could strike from deep with threes—he possessed a beautiful stroke—and he shot a higher percentage from outside the line than Duke's JJ Redick. He could bull-rush the basket with muscular authority, finishing with a dunk or a layup. He even owned that relative rarity in college basketball these days: a midrange game, pulling up easily off the dribble for jumpers. His offensive ability was off the charts, reminiscent at times of the Seattle Supersonics' Ray Allen, and college coaches and NBA scouts knew it.

He also displayed a knack for late-game heroics. Against Connecticut in January 2004, McCants came off a pick in Carolina's Long Beach set and buried a three-pointer to beat the Huskies, then the number one team in America. It didn't always work, though: In the final moments of the second Duke game against JJ Redick last spring, he'd tried to step back for a three and lost the ball off his leg. Redick threw himself onto the floor and grabbed the loose ball. Game.

Perhaps impelled by the memory of times like this, McCants prepared for the closing moments of his games like a graduate student in film studies, going image by image through videotapes of his favorite players at the ends of their tight games. "I watch those clips," he explained, "and I don't watch the moves of the players. I watch their *faces*. I study the body language, how they wipe the sweat off. Michael Jordan, for instance—he's so calm when he wipes the sweat off his face." Was it any wonder that a player studying the visual evidence of triumphant psyches, the tics of their faces, the way they dried their sweat—was it odd that such a player might be exquisitely sensitive in his own inner self? And that he might be trying to achieve an equanimity in line with his surpassing skills, hunting for clues in the expressions of his heroes?

Like Jordan, McCants was a North Carolina boy, born and raised in Asheville, Roy Williams's hometown in the mountains. His father wrote, "4-12-86 next Michael Jordan" in McCants's baby book. Like many in-state kids, he rushed out at the halftime of UNC broadcasts, pretending to be Michael Jordan sinking a last shot against Duke. Unlike all of the other North Carolina kids playing out similar scenarios, McCants demonstrated such a precocious gift for the game that it appeared he might actually be in a position to take those last shots one day. He starred at Erwin High School in Asheville, then transferred for two seasons to New Hampton Academy Prep in New Hampshire, where he finished his high school career as a McDonald's All-American, averaging 25 points a game and seven rebounds.

The conventional wisdom held that he had flirted with Duke only in order to hasten an offer from his childhood love, North Carolina, but McCants told me that he had actually seriously considered becoming a Blue Devil. In the summer following their sophomore years in high school, he had discussed doing exactly that with JJ Redick, who would eventually be his counterpart at Duke. McCants said: "I told him, 'You know, we could both get scholarships there.'"

The thought of McCants shooting jumpers and slashing to the basket as a Blue Devil was disconcerting, to say the least, especially given that he was a North Carolina native brought up under the star of Michael Jordan. In the end, the Asheville native succumbed to North Carolina coach Matt Doherty's skillful pursuit. It was a result that both parties would eventually have cause to regret.

"I'm probably the most private person on the team," he once told me. "I just keep to myself. I don't do anything out of the ordinary. Just stay in my room." His shyness dissolved most fully on the court. He had that performer's tendency for boldness in front of thousands, introversion offstage.

For McCants, basketball had always functioned as a release, "a stress reliever," as he put it. "Whenever I'm feeling depressed, or just need to get away, I go to the gym." He would visit the Dean Dome at any hour of the day or night, throwing tennis balls for hand-eye coordination, shooting jumpers, just dribbling.

So when his basketball life began to disintegrate during his freshman year under Matt Doherty, it shouldn't have been surprising that McCants himself began to crumble. He had started his freshman year on fire, scoring 28 points in his first game on 11 of 14 shooting, three of three from beyond the arc, seven rebounds, and two steals. It was the most auspicious debut of any UNC freshman ever. He followed that a couple of weeks later by being named the MVP of the preseason NIT.

Despite McCants's offensive brilliance, he tended to turn it on and off on the defensive end, an inconsistency unappreciated by the hard-nosed Doherty. The coach cut back McCants's minutes, leaving him on the bench for long periods for the first time in his basketball life. And when Doherty complained to the media about his freshman's defensive deficiencies, McCants resented the public humiliation. "He brought up my defensive effort in the press. He put the perception out there that if I didn't play defense, I was a bad kid. Once he benched me, I was a bad kid. And there was nothing I could do.

"I started out, I was getting the ball fed to me," he said. "So once the ball was taken out of my hand, I didn't know how to react. I felt like I could make a big impact my freshman year, and I did in the beginning. But once the controversy started building up, it hit me pretty hard."

At the practice following an away game at Georgia Tech in late January 2003, McCants wept. "I cried because I thought that there was nothing I could do," he said. "I had just played one of the worst games that I've ever played. I was stressing."

For the first time in his life, basketball no longer gave him pleasure. His refuge from stress had turned into the source of his unhappiness. He wanted to quit. "It was a struggle every day I went into the gym," he said. "I grew up loving the game, loving Carolina, and I got here, and I was like, I didn't know it was going to be like this."

All his life he'd been praised for basketball, for his intensity on the court. He even led the team in its pregame circle, psyching up his teammates with chants and woofing. Suddenly, the very floor of his esteem had been cut from under him by his own coach. "When a coach establishes that type of relationship, it's not a choice for a player," he said. "I just wanted to survive the rest of the season and finish up strong, for my reputation's sake and the team's sake. But the trust was kind of shattered."

Like the freshmen of the season before—Melvin Scott, Jawad Williams, and Jackie Manuel—McCants spent time at the home of the team's academic advisor, Burgess McSwain. A silver-haired lady who lived with her father, favored gaudy hats, and doted on her dogs, she had served as the team's academic counselor for over 20 years. The sociologist Thad Williamson, a longtime admirer of the Carolina basketball program, tells a story about how McSwain stayed on top of her charges even when they returned to finish their degrees after going pro early, as in the case of Michael Jordan. She'd heard that Jordan, already an NBA star and back in town for summer school, was playing golf when he should have been studying. She drove to Finley Golf Course, to the tee where Jordan was preparing to swing, rolled down her car window, and told the player in no uncertain terms to get his ass into the car. He obliged.

Her tough love proved invaluable to the players during the tumultuous Matt Doherty regime. In the process of helping them with their studies, counseling them about the basketball program, giving them advice about girls, and letting them fall asleep on the couch at her home, she also provided her boys with compassion and steadiness, two qualities in short supply during Doherty's tenure. For that, she earned the team's gratitude. When McSwain died of cancer in 2004, the players served as pallbearers at her funeral. For the current season, McCants had inscribed "Burgess RIP" on his left Nike sneaker.

He had been studying at McSwain's at one point during his freshman year when Matt Doherty came by, seeking to have a conversation. When the coach showed up at the front door, McCants went out the back. In another instance, Doherty tried to set up a meeting between McCants and a psychologist who had a long relation with the North Carolina basketball program. McCants felt stigmatized, betrayed.

He tried to block out the negative energy. "I took it upon myself," he said, "that there was nothing the coaches could say that could bring me up or bring me down anymore."

McCants kept abreast of the battering he was taking in the media. He routinely read *Inside Carolina*. "It gives you a good sense," he said, "of how people really view you. You're not just reading people who praise you; you're reading people who are bringing you down all the way. And you think that they're the same people who are asking for your autograph when you come out of the tunnel at the Smith Center. And they're the same people who are saying you can't hit the fifteen-foot jump shot. Some people really think they know a lot—who never played basketball."

McCants invested the press with considerable power. "They can't make your personality," he said, "but they can make people judge you, determine how they look at you."

He rarely made attempts to placate reporters, however. As fascinating as it was to watch Rashad McCants play the game of basketball, it was at least equally compelling to watch him tangle with the press in the players' lounge after the games. He froze reporters just by arching an eyebrow shadowed under the brim of his cap. His eyebrow quietly said, "How could you have asked such a stupid question?" Once I witnessed a beat reporter request that McCants recap a play from the game that had just ended. Standard sportswriter stuff. McCants arched his eyebrow. "You were there," he said. A couple of beats passed. "You saw it." He shrugged. What was there to recap? The reporter wrote down McCants's meager response, but I don't think I saw it in the next morning's paper.

The list of questions that his eyebrow disliked was long indeed. How will you prepare for _____? How will you stop _____? Do you feel your defense has caught up to your offense? Then there was the worst question of all, asked time and again ever since his sophomore season: How are you and Coach Williams getting along now?

That question not only raised an eyebrow, it made McCants frown, roll his eyes, stare at the floor, stare at the interviewer, smile in disbelief, smile in confirmation of his worst expectations, shrug, shake his head, and mumble. Or say absolutely nothing at all.

But when he did open his mouth and deign to give an answer, McCants defied the old Carolina custom of uttering pabulum, of trotting out conventional pieties celebrating the sun, the moon, and getting back on defense. At the beginning of the season, for instance, he'd compared playing at North Carolina to being in prison. "I'm in jail right now," he said. The remark had been taken out of context and billboarded by a local television station: Tar Heel Prison! McCants at 11! What he actually meant was interestingly complicated and astute—that life as a college basketball player tended to be regimented and that the spotlight (from the television cameras? or the prison tower?) glowed uncomfortably hot on a young man, especially one who valued his privacy, perhaps because he was shy.

On the other hand, McCants had indeed said that being at Carolina was like being in prison, an observation that in an age of public relations and shock headlines did not get parsed too finely. Roy Williams worried publicly about the effect the quip might have on recruiting. McCants went into the doghouse.

He'd been there before. Against UNC–Wilmington in December 2003, McCants and his teammate Jesse Holley had been banished to the locker room for not having stood and cheered another teammate's defensive play. "If you guys can't cheer, get your asses to the locker room," Williams told them. Then there'd been his infamous performance—four points, five turnovers, lax defense—two games later against Kentucky, for which Roy Williams had benched him, angering McCants. "When I was playing bad," McCants said, "he told me he could do without me. He said to watch. He said I just had to learn from it."

"At the time," McCants said, "I felt like I wasn't given a fair chance to really get into that game all the way. I was pulled in and out sporadically throughout the game. I did turn the ball over a couple of times, and I wasn't really aggressive. But I definitely didn't get the opportunity to open it up. Some players need a bucket or a steal to really get 'em going."

He initiated a conversation with Williams after the game. He apologized for his play, committed to improving his attitude, and told the coach that he needed Williams to show faith in him. Williams accepted his star's pledges, and added, "If you stay all four seasons, then I'll be in the green room with you at the NBA draft." He'd never promised that to another in the distinguished line of players he had instructed at Kansas—not Paul Pierce, Nick Collison, or Kirk Hinrich. Their relationship improved. "He knows I care about him," Williams said. Eventually, the coach would even defend McCants to the media hordes, saying he'd never had a player so scrutinized, whose every move was dissected like a politician caught in a scandal.

There had still been bumps. Against Maryland in Chapel Hill in 2004, McCants attempted a showboating, double-clutch, from-the-hip dunk only to have it catch harmlessly on the rim. His coach fumed. But the next day at practice, I witnessed an exchange that indicated an increasing tolerance between the two. "I'm going to give you another chance to do that dunk," Williams told McCants.

McCants backed up a couple of steps, then sailed in for the jam, this time throwing it down with the shrugging elegance of an aristocrat. The coach laughed and patted Rashad on the butt. "All right," he said.

"Coach was mad because he didn't think I could make it," McCants later said. "I was winded during the game, and I didn't have my legs. But I could do it. And I did it. Like I said I could do it."

For all of his alternating bouts of silence and free speech (and it was the vacillation between them that really threw everybody off, his teammates included), McCants appeared to have made peace with his coach and his team this season. On the way back from the U.S. junior-team trials in July 2004, Sean May had taken it upon himself to counsel McCants in the basics of media relations. "Don't be unapproachable," he told his teammate. "Smile if somebody comes up to you. . . . I'm your boy and I'm going to tell you: There's stuff you need to change." Tellingly, May had made the team, and McCants had been cut, having cruised over the last couple of days of tryouts after dominating the initial sessions. Kelvin Sampson, the Oklahoma coach who cut him, said that McCants was "without question the best player at that camp."

"Sometimes, you've got to play off the world's politics," McCants said. "You got to learn how to play a part. Sometimes people want you to smile when you're not really happy." McCants was learning that the brooding face he presented to the world caused more people to inquire about his inner state. He was belatedly discovering that sometimes a smile was the best disguise.

On the court, he'd agreed to keep his solo artistry to a minimum— shoot a little less, drive a little more, and pass the ball around. Having served so dutifully as a media whipping boy, the change in the McCants narrative had broadcasters like Mike Patrick greeting his low-scoring games as a sign of maturity, whereas the same stats in previous seasons would have been cause for attack.

And indeed, while his scoring average had dropped from 21 points a game to right around 16, his assists had gone up; by midseason, he had almost surpassed his total of the previous year. He was doing more by doing less.

THE DOG DAYS OF WINTER
UNC versus Miami
The Dean Dome
January 22, 2005
8:00 P.M.

The winter winds blew in from Canada or Virginia or Carrboro or some such place; freezing rains clattered to earth and coated the town with a nasty glaze of ice. Like a mass of refugees on skates, my sister, Annie, and I and thousands of Carolina fans slipped and slid our way down Skipper Bowles Drive and up the treacherous slope to the Dean Dome, where we belatedly took our seats just before tip-off as if we were typically late-arriving Ram's Club members. Down on the floor, Rashad McCants was going through his usual jumping-jack routine just before the tap, intently bouncing to "Jump Around," the 1992 hip-hop hit by House of Pain. (The University of Wisconsin proudly claimed first use of this number at a Badgers football game, and credited it with arousing an entire stadium. Of such did universities now brag.)

What I loved best about this moment, however, were the old ladies in the Carolina-blue cashmere sweaters, Tar Heels brooches pinned on near the collar, who joined the hip-hop nation to the best of their ability, bobbing and whoo-whooing and putting their hands in the air like they just didn't care. They inspired me: They gave me an old age for which to shoot.

But all of this pregame jumping may have been the night's most energetic moment. The Tar Heels, perhaps underestimating the competition, appeared sluggish early. When Melvin entered the game at 15:11, the score was 15 to 11 in favor of Carolina. Melvin made a beautiful drive after receiving a pass from the freshman point guard Quentin Thomas, hit the layup, and got fouled. However, he missed the free throw. And then missed a deep three a play later. Melvin out at 10:12.

With eight minutes to go in the first half, North Carolina was leading by only six, 21 to 15. They were becalmed in the Sargasso Sea. The boat rocked in the waves; the fans were lulled asleep. Given the talent discrepancy, the lead by now should have been much greater. Melvin reentered the game at 5:58 with his team up seven and again took a great pass from QT, pushed hard to the basket, and got fouled. This time, he swished the free throws. UNC inched its way to a 14-point margin at the half, 40 to 26.

The second half began with May scoring on two successive fast breaks coming off missed Miami shots, and before the Hurricanes could adjust, they were down 60 to 36. Hardly six minutes had gone by. Melvin was passing up threes, even when wide open. He hustled back on defense. He was trying to do things the way he'd been asked.

There was only one problem with North Carolina's performance: They were slinging the ball all over the court, just as they had been in practice on Thursday. With 8:38 left in the game, Roy Williams called time-out, but rather than attend the huddle, he slammed his clipboard onto the floor in front of the players and stalked off to sit by himself on the bench. This was a new type of tantrum, and I liked it. He sat there at a distance from the team, legs crossed, fuming. "He looks just like Daddy," Annie said.

Assistant coach Steve Robinson took over where Roy left off and seemed to be ripping the team. When the players returned to the court, their play perked up, and they finished with a 20-point win, 87 to 67. It was a workmanlike victory, not especially impressive, and yet to win by 20 when your game was unsteady might have been a sign of good things to come.

In the press conference, reporters asked Roy Williams about his burst of anger. "I wasn't mad about the economy," he said, "just the way we were playing. Ray Charles could have seen that. I was about to explode. I was like an alien creature over there. Sometimes it's better not to say anything than to say something."

On cue, he was silent for a moment. The cameras clicked and whirred. "It's the dog days of the season," he said. "And we're taking chances with the ball."

In the players' lounge, the guys had more to say about Roy's silent treatment than the game. "I've seen him break a couple of clipboards," Jackie Manuel said, "but never not say anything."

"I'm not going to give you any dirt," Rashad McCants told the press.

"I knew when he called time-out, something was about to happen," Sean May said. "Break a clipboard or something."

"We deserved it," Jawad Williams said.

Melvin left the locker room wearing a Joe Montana jersey and cap. A little blond-haired boy saw him and ran toward him, calling excitedly, "Marvin, will you sign my shirt?"

Melvin said, "You want me to sign Marvin?"

"Oh, I mean Melvin," the little boy said, embarrassed. It was that kind of night for Melvin. He was grateful to be playing. And he'd played hard tonight, driven to the hoop, taken the contact. Like Charles said he should. But things weren't going quite as he had hoped for his senior season.

MELVIN IN DESPAIR

The Wake Forest game, two weeks before, had done a number on Melvin's confidence. The previous year, he had started the game at Joel, scoring 14 points in the first half and 20 for the game, and the Tar Heels had won on the road in the ACC for one of only two times that season. "Everybody was like, Man, you hit those five straight shots and you're the reason we won that game." This year, he said with a downcast and mystified air, "I don't start, play about three minutes, and we lose."

We had come to the Outback Steakhouse for supper. After the waitress brought Melvin's Shrimp on the Barbie, he bent his head and put his hands together in prayer. When he finished praying, he studied his entrée and picked up what seemed to be a blackened piece of toast, turning it this way and that, trying to figure out what it was.

"After the Wake game this year," Melvin said, "I was crushed. And my mom was crying."

And to make it worse, Roy Williams had stopped Melvin in the tunnel after the contest and upbraided him. "You know, Melvin," he had told him, "with attitudes like yours, we're not going to be a good team." And then, to make matters even worse, Williams had shown the film of the game and enumerated Melvin's failings. "He showed my first shot where I took a bad shot. I thought I would make it. I guess he was just frustrated because a lot of guys in that game were taking bad shots, not going for loose balls. I guess he decided to take it out on me a little."

Melvin looked as somber as I'd ever seen him, near defeat. A misery radiated from him that seemed to shut out the happy noise roaring around us. He'd spent the summer working at his point guard skills. He'd dribbled and dimed, he'd exercised his conscience, he'd passed up shots for the team's sake, even in pickup games. Now he was playing less than he had last season.

"How do you deal with these reversals and frustrations?" I asked Melvin.

"It's tough, man," Melvin answered. "I had a sit-down with the coach the night before Wake, and he explained what he wanted. I didn't really understand but I . . ." His voice trailed off.

"What did he say?"

"I was like, 'What do I do now to get more playing time?' He was just telling me to get into the game more, to lose myself."

"Lose yourself in the game?"

"Yeah. So I'm like, okay, I can take that. Nothing wrong with that. He said that if I miss a shot, I don't play defense or I whine about shots. But losing yourself in the game when you're a shooter means that if you see a shot you can knock down, take it. Be aggressive. Not necessarily shooting every shot, but be aggressive. If I'm showing the defender the ball and being aggressive, other things open up for my teammates and myself."

There were other demands. "He told me that during pregame, instead of joking around so much, lean back some and don't joke so much. So now when he walks past, I want him to see that I'm focused and serious. Of course, when I'm having fun, that's when I'm playing my best. But, whatever . . ."

Melvin moved his food from one position to another. It looked like he was diagramming plays with the shrimps, coaching them. "I know we're a better team this year because we're unselfish, but I just sit around."

"It's hard, man. I don't talk to anybody here about my problems." He didn't want to confide in his roommate, the forward Jawad Williams, because Jawad was having a great season and didn't like Melvin complaining about their coach. "My mom says, 'Just stay strong.' She's hurting more than I am."

As for his brother, Charles, his advice to Melvin was simple. "I had told Charles that I only have a certain amount of time to do everything when I was out there, and Charles said, 'Well, hit every shot.' That's not fair. I can't hit every shot!"

"You've got too much pressure riding on every shot," I said.

"Yeah, and when I think about a shot," Melvin said, "I miss."

Melvin was eating slowly now, picking at the shrimp with the deliberation of a diner for whom food had lost its taste. The only thing that offered real sustenance right now, he said, was his religious faith. "Without that, I would just let go," he said. "I wouldn't even be playing basketball right now. There's a saying in the Bible: He who humbles himself will be exalted. I just got to be humble and patient and the Lord will lift me up in due time. I don't think it's fair what I'm going through right now, but that's life. Life isn't fair."

In middle age, I had become reluctantly aware of that, though I contemplated the degree to which I had brought misfortune on myself. Either way, though, it seemed that life lacked justice. I wasn't sure whether it was good or not to accept the proposition that life wasn't fair. In some ways, it seemed the beginning of wisdom. In other ways, it seemed a capitulation to the sorts of injustice that only human beings could rectify.

In the view of most Carolina fans, however, Melvin's plight was just. He was a shooting guard who didn't shoot that well. He didn't handle the ball skillfully enough to play point. And the team was functioning beautifully in its present configuration. Yesterday's hope—as Melvin and all the players had been, highly recruited as they all were—had become today's disillusionment, a 22-year-old who sat on a bench as much as an old guy in a park.

Melvin considered the ironies of his situation. "If I sit around and pout and complain," he said, "everybody will say, Melvin is a bad guy. North Carolina is winning and number three in the country. But when we lost to Wake, I almost lost it myself. Because we *didn't* win, and I'm still on the bench."

In the practice gym the next day, Melvin decided to do some extra shooting. He had come to college renowned for his long-distance marksmanship, and now he was working hard to reclaim his old reputation. Today, he started out by banging in his customary three bank shots from each side of the basket, about three or four feet away. He did this to get a rhythm going. "I'd be a better shooter if Smitty were around," he said, referring to Meredith Smith, his high school coach at Southern, now an assistant coach at Maryland Eastern State. In college, Melvin had never quite found that older mentor that had saved him in middle and high school.

The way Melvin described it, Dr. Andrey Bundley, his middle school principal at Greenspring, "was the cake mix and Smitty put the icing on it." Melvin, of course, was the finished cake. "Smitty would say, 'I know I'm crazy, but Melvin is even crazier than I am,'" Melvin said with pride. "Not all guys were crazy enough to understand Smitty. They weren't on that same channel. Without Smitty, I might be in college playing basketball, but not at North Carolina. Mentally and physically, Smitty prepared me for college. He made me believe. College coaches give you love, but not the type of love a high school coach gives you."

Melvin believed that when he got to North Carolina, the coaches hadn't really understood who he was. "They thought I was a con man," he said. "Shifty." Not Smitty: He had seen Melvin in the favorable light of their mutual dream. And thus, Melvin pined. When times were toughest, his thoughts drifted back the most.

At the free-throw line, Melvin positioned himself carefully. "Smitty told me every free-throw line has a nail in it, dead center," he said. "Find that nail. Put your right foot on it, if you're a right-handed shooter. Put your left a little back. I like to take three dribbles and spin it. Then I take one more dribble. Spin the ball again. And shoot."

By design, Melvin didn't bend his knees as much as some players at the line. He held the ball a little to the right as he came up in his shooting motion, his elbow tucked in. And now he nailed every shot. "Smitty told me you should never miss left or right if you're shooting with the right mechanics. Only off the front of the rim or long."

As he went about his drill, he channeled Smitty's teaching. "Aim for the back," he instructed himself. "If I hit the rim and it goes in, it doesn't count."

More than anything else, a shooter needs confidence, Melvin declared. He needs to put the shot up without thinking, without worrying that if he misses, he's coming out of the game. If he's a shooter, he needs to be able to pop a jumper even over the outstretched arms of an onrushing defender. If he's a shooter, he should be able to take the occasional bad shot because—if he's a shooter—he will often hit it. All of which put Melvin in mind of his counterpart eight miles away at Duke, JJ Redick. Last year, they'd both been the starting shooting guards for their schools. Now only Redick enjoyed the distinction. "Sometimes I envy JJ Redick," he confessed. "I mean, I'm happy to be here and all. But just put my face on JJ's body. I can get my shot off, but he has those extra inches." Melvin was listed at six-one, which may have been generous; Redick was an easy six-four, six-five. Melvin pointed out that when JJ's shot had gone south the previous season, "the coach stuck with him." "I watch him," Melvin said, "and it's like, man, that's me out there. If I could just have shots like that. I love watching him play."

It wasn't just Redick's height—and playing time—of which Melvin was jealous. "One thing I didn't look at in coming to the Tar Heels was the style of play," Melvin said. "We go inside-out." He meant that North Carolina traditionally looked first to the big men near the basket for high-percentage shots before pitching the ball out to the guards for long-range jumpers. "I tell young players now to look at style of play," he said. "Like Jason Williams—at Duke, he played the point, and he got to come off ball screens and either shoot or hit somebody cutting. That's what I did in high school. That's my game. And coming off screens like JJ does, oh, man . . ." Melvin smiled and shook his head hungrily, ruefully.

"I love watching those guys play, actually," he said of Duke. "Especially on offense. They run a more perimeter-oriented system." Here at

North Carolina, Melvin had concluded that however much he wanted to play like Redick, he was not going to be allowed that privilege. "I'd rather play more," he said, "than take five quick shots and sit down."

He hadn't chosen North Carolina because of its style of play, Melvin said. He had chosen North Carolina because nobody in Baltimore thought he'd make it here. And yet, he had. His prayers had been answered.

THE BEST SUMO WRESTLER AT CANYON RANCH SPA

I was the best player in my neighborhood, which is a little like saying I was the best curler in Trinidad, or the best sumo wrestler at Canyon Ranch Spa. My neighborhood was not exactly Rucker Park in Harlem. Kids tended to drift in and out of games depending on the lesson schedule for the day. We carved our first court out of a hillside overlooking a creek bed that ran below our house. Rocks and exposed roots made every dribble an illustration of the chaos theory in physics. The ball ricocheted at unexpected angles, like a neutrino, rarely returning to your palm with any sense of homecoming.

If you missed a shot badly, or sometimes even hit it so cleanly that it dropped through the net as unimpeded as an air ball, there was always the likelihood that the ball would disappear down the hill. Down, down, down, it went. A rule evolved (natural selection) that the worst player among us, the one we felt was lucky even to be playing, had to retrieve the ball. Often, he volunteered for the job. "Ball!" we'd yell and down he'd go, churning through the branches and brambles like a hound chasing a fox. (I will not name him, because he is now a high-priced lawyer who might sue my ass.) "Kick it," we'd yell maliciously once he reached bottom and ball. This resulted in his trying to kick the ball back up the hill, which only meant that it would bound off more trunks and branches and rocks and roots and roll back to the bottom of the hill, only to irritate our retriever into kicking again, harder and madder. And so it went.

Having engineered the hoop and backboard about ten feet up the side of a slender young maple tree, my father at first felt some proprietary interest in the games. He would join us, wearing his Saturday-afternoon work khakis and a torn brown sweater, unraveling at the cuffs. He had played on his high school team in Huntersville and still owned a Forties-style game, stiff, straight up, staring at the ball while dribbling like a man mesmerized by a bouncing sun. It took him a long time to make his move, which usually involved dribbling slowly across the extruded surface of the court, bucking his shoulder into the 10- or 11-year-old defender, who stood about a foot or so shorter than he did, then pushing the ball onto the backboard with two hands and waiting patiently for it to drop through the rim. If it did, he would exclaim something that sounded like "OOOMPAH."

If I may translate for my dear departed dad, I think that was his way of saying, "In yo' face, son."

PRACTICE
The Dean Dome
January 27, 2005
4:00 P.M.

At practice today, Melvin's body language was depressive. He sat by himself next to a rack of towels and jerseys. He was either the life of the party or a loner. Now he stalked the end line behind the assistant coach Steve Robinson, who was officiating.

He had entered the play as part of the white team—the starters—but missed a three to the side of the rim. He had measured the shot too carefully. He appeared to be thinking too much. Then he suffered a cut, so he left the court once again to be attended to by the trainer, who wrapped a bandage around his elbow. He looked detached from the team, as if he had become the shooting guard no one needed or wanted. He lingered on the sidelines by himself, waiting to be called into the action.

He couldn't have felt much better when he went back into the game, because Rashad McCants dropped a clean three on him, despite his avid defense. McCants had been putting on an exhibition all practice, scoring with a dispassionate ease—threes, dunks, midrange jumpers. He had turned the Smith Center into his backyard.

For Melvin, it was the opposite story. With the white team running offensive drills, he missed two threes in a row and spat on the floor by the basket support, looking frustrated. As the rest of the team took a break for water, Melvin sat by himself next to the towel and jersey rack. Then, while the rest of the team relaxed in a long row of folding chairs, he shot jumpers, the ball fed to him by a manager.

When practice resumed, Roy Williams appeared a little gloomy himself. As the blue team fast-breaked towards the basket, Melvin called for Quentin Thomas to stop the ball. Neither did. "Stop the ball!" Williams yelled, "so we don't give up a frickin' layup."

On the next play, the white team lost the ball. Williams screamed for the white-team subs to move faster to the side of the court. "I'm tired of wasting our frickin' time," he complained.

With Jesse Holley, Wes Miller, and Charlie Everett leading the way, the blue team was shredding the starters. When Holley stripped Sean May and glided in for a layup, Williams went apoplectic.

"STOP THE BALL!" he shouted, stomping on the court.

And then, almost instantly, things got better, at least for Melvin. Denied an initial entry pass into the post, he dribbled away from the defender, spun suddenly, and bounced the ball into Jawad Williams. Roy Williams stopped the action. "Guys, what Melvin did there was great," he said.

That was all Melvin needed to hear. He responded to praise as if he'd just received a transfusion of Gatorade. You couldn't help but wonder whether his coach had sensed his flagging spirits and tried to pump him up. That's the way it had always been for Melvin: Express belief in him and he would get up at five in the morning, run extra suicide drills, fly around the court, clap for his teammates. Ignore him and he died inside.

Reenergized, Melvin traded threes with David Noel, who had been hot all day from the outside. Back on the sidelines, he had Noel and QT in stitches about something. At the top of the foul line, Roy Williams was insisting: "Everybody must understand, we will not foul a jump shooter."

As the team finished practice with its customary running, the ever good-natured Marvin Williams laughed as he ran behind Sean May on the last 22, then caught him at the end and pretended to break the tape as he crossed the end line.

THE BEAT-DOWN
UNC VS. UVA
University Hall
Charlottesville, Va.
January 29, 2005
12:00 P.M.

Sean Singletary opened the action by downing a three-pointer for the Cavaliers. That was about it for UVA highlights. They fell behind, 11 to 3, and it got worse from there. With North Carolina up by 36 at the half, and by as many as 50 late in the game, there wasn't much suspense to this one. Except for my mother. She was worried that if North Carolina kept fouling, Virginia would catch up.

"Mama," I said, "we have a thirty-six-point lead."

"I know, but I've seen us lose a twenty-point lead," she said.

"Mama, they're not going to lose this game. Trust me."

"I'm glad you're so confident."

I found myself inadvertently slipping into my father's role of assuring my mother a Tar Heel victory. The difference was that he liked to promise them in close games, seesaw battles that grayed the hair. I was doing my best to ease her anxiety in a contest that North Carolina would end up winning 110 to 76.

"I think we're going to make it," I told her.

"Hush," she said.

When it was all over, my mother said, "That wasn't too bad."

THE SERPENT AND THE SLEEPING MAN

That Sunday, the journalist put on a coat and tie and departed to attend church with Melvin Scott, leaving the beast behind, presumably sitting on the couch in raggedy gym ensemble, poring over the sports section (his Sunday Bible) while sipping coffee. In his eyes, contemporary Christianity had about as much to do with Christ as—what was that famous phrase from the women's movement?—a fish did with a bicycle.

The journalist tried to keep an open mind about such things but even he was feeling a little swamped by voluble, proud believers in these evangelical days—the end times, if you believed some of the brethren. Faced with all of these folks who just couldn't wait for Christ to return on the Day of Judgment and kill a lot of people who weren't like them, he found himself increasingly sympathetic to the reticent, scholarly Presbyterians he had grown up around.

In regard to Christianity, Melvin Scott had no doubts about its truth or its power. He had started attending church in his early teens with Dr. Bundley back in Baltimore. He couldn't explain what exactly happened or why, but the experience had changed his life. His mother, his brother, Charles—they all marveled at the sudden transformation in Melvin. "I guess God just came into his life," Charles said. Melvin began carrying a Bible around with him. He didn't like cursing. It appeared that Christianity was a supreme method of self-discipline for him. "My temper doesn't overcome me now," he explained. "Recently, I watched *The Passion of the Christ,* and if He can go through that, then I can restrain myself. What is happiness if you can have God through suffering?"

King's Park, the church he attended on Sundays and occasionally during the week, lay to the east of Chapel Hill, several miles off Highway 54 on the way to Raleigh. Massive and modern, the church rose like a sports dome at the edge of a gigantic parking lot. The whole complex was situated deep within a development that had been hacked out of scrub woods, one of those awkward American zones of in-between: part office park, part mall, and part suburb. And yet to newcomers in the area, of which there were many, the space would have felt at least as familiar as an interstate rest stop.

Inside, an enormous and nearly filled sanctuary yawned before me, looking more like a multiplex theater than the weatherworn Baptist churches that still dotted the countryside. The place was outfitted to provide as much entertainment value as any movie palace.

I sank into a plush theater-style seat that should have had a drink holder in the arm. On a stage below, nine singers swayed back and forth, dressed in black dusters, belting out a pallid rock number about Jesus. Above the singers' heads, two video screens proclaimed the song's lyrics in flashing lights: "Jesus, you are altogether wonderful." The singers gestured for the audience (for that is what it was) to sing along. The song struck me as insipid kindergarten sentiments for big, bad adults. I grew instantly nostalgic for tattered hymnals shared in a pew. Archaic words, majestic organ. Give this agnostic grandeur and awe or give me nothing at all.

Melvin arrived, chewing gum, wearing blue warm-ups, just as a bald black preacher was shouting, "Lift up your hands!" Around us, white and black families raised their hands in the air, as if signaling a touchdown. Several rows beneath us, we noticed Jawad Williams, his hands held high.

"Sorry, man," Melvin said. "I slept late." He slipped ten dollars into the collection envelope. On the envelope was written: "A generous man will prosper; he who refreshes others will himself be refreshed. Proverbs 11:25."

Down on the stage, a white minister asked the congregation, "Anybody know about financial famine?"

"Oh, yeah!" somebody yelled. He sounded happy about it.

"A financial famine is when you open the refrigerator and see a stick of margarine," the preacher said.

Melvin hunched forward in his seat, listening intently. Several young members of the congregation turned to smile at him. He nodded in a friendly manner but remained focused on the service. The minister was telling how he had once bought a car on credit without knowing that his congregation had secretly purchased an automobile for him as a gift. He had lacked sufficient faith. "God can even bring forth produce when you are unemployed," he said. "God is El Supremo, not El Cheapo."

Now it was on to the main event. In front of the congregation stood a middle-aged fellow in a suit that constrained a vigorous bulk. He lived now in Nashville, Tennessee, but he used to be minister at King's Park. During breakfast this very morning, he'd experienced a vision. "A man was resting in a glade," the preacher said, "but a serpent came, and it was so long that it was amazing." He took a long swig from a bottle of water. He was already sweating.

The minister stopped, drank again. "I don't like to share things like this," he said, "but the snake began to coil around the man, slowly and imperceptibly. And unusually for me, my head began to get very tight as I envisioned this, and a pain began to come, and the Holy Ghost said to me, 'He wants to crush *your* head, as that snake wants to crush the powerful man's head.' The Word of the Lord began to wake up the man against the snake and break the restriction. Muscles are built to struggle. The more you struggle, the bigger they get."

"Amen," said Melvin and the rest of the congregation. This was a hell of a sermon. Newfangled church, but old-style preaching, alarmed by the ubiquity of sin. The minister worked at allusive angles to the Biblical material, drawing his conclusions from the Word and his own humbled life. He seemed to have felt these things in his heart. He didn't appear overly virtuous, and that was a virtue in a preacher.

He described another sleeper, Samson, seduced through carnality and compromise. He had been a mighty man born to shatter the Philistines. Delilah seduced him. He had let her in on the secret of his strength. "If there's a story about how blind sin makes you, it's this one," he said.

"Sampson jumped up from his sleep, shaven-headed," the preacher said, "and he was thinking, *I've played on the Internet before and it's never gotten me!*" The congregation rippled with laughter. "You can't shake off sin like you used to!" the pastor suddenly roared. "While you were slowly seduced into a moral slumber, you lost your connection with God."

The preacher said, "Some of you are just trying to sleep through your sorrow, sleep through your pain. The Lord says, 'Wake up and eat. A fire is burning in front of me, the fresh hot bread of Heaven is here.'"

He lowered his voice to that of a seducer's. "Wake up," he whispered to the church. "Because even now you're in the jaws of distress."

He invited the congregation to come down to the altar and receive the Bread of Heaven. Melvin and Jawad arose and went to join a cluster of worshipers, kneeling and praying with great intensity. "You don't have to tear the bread," said the minister. "Take the whole loaf!" Jawad soon returned to his seat, but Melvin stayed at the altar, on his knees, bent over in prayer until no one else was left. When he finally returned, he was wiping tears from his eyes.

"There have been divine land mines planted here and they are going to explode," promised a minister. "Watch me, watch me," he said, reaching out toward the congregation. Melvin stretched his hands out in response. And me, I stretched out my hands, too. I don't know why. Maybe I just believed in being polite.

IT'S NOT FOOTBALL
Duke versus Virginia Tech
Cameron Indoor Stadium
January 30, 2005
8:00 P.M.

That evening, several hours after the conclusion of church, it was time for Duke to play Virginia Tech. In an abiding spirit of Christian love and compassion so appropriate to the day, my mother exclaimed, "They make me sick." She had just watched a gaggle of Blue Devils bumping chests after Shelden Williams had rebounded a Shavlik Randolph miss and been fouled.

We were scouting Duke, if that's what you want to call it, my mother on her sofa, me on mine. The Blue Devils and the Tar Heels were coming up in just about a week and we needed to get another look. Virginia Tech, Duke's opponent for the night, had been surprisingly impressive in its first ACC season, having won four straight conference games, giving my mother and me reason to believe that they might play the Blue Devils tough.

"If they're going to call all those fouls on Virginia Tech," my mother said, "I hope they'll do the same against Duke."

"Mama, you might as well ask for the moon."

"How do they get away with it?" she asked.

"I'm sure scientists are studying that as we speak," I said.

She shook her head dolefully. "That would be nice," she said.

Her friend Neal Campbell had recently sent a press release issued the previous week that we had somehow missed. I couldn't vouch for its authenticity, but it seemed apt.

John Swofford announced today that a foul is tentatively scheduled to be called against Duke some time in the first half of their game against North Carolina State in Raleigh on Thursday, January 13. In a joint press conference with Duke head coach Mike Krzyzewski, NC State head coach Herb Sendek, and ACC director of officials Tommy Hunt, Swofford said an agreement had been reached for a touch foul to be whistled on an as-yet-to-be-determined Blue Devil player around the 7-minute mark during the first half of the game at the RBC Center. "We are very excited to arrange something that hasn't been seen in our conference since 1978," said Swofford. "I want to personally commend Mike Krzyzewski for agreeing to this unconventional deal. We all know how reluctant he has been to allow any calls to go against his team."

Krzyzewski insisted that this move was purely a gesture of generosity aimed at rehabilitating his public image in light of his recent lip-synching fiasco during the Clemson game.

"Mike is really being a good sport about this," noted Sendek. "This is a once-in-a-lifetime opportunity to play them with a foul being called against their squad. To have it happen at home will be especially enjoyable for our fans."

Tommy Hunt said that he did not expect any more fouls to be called against Duke this season but did not rule out the possibility of one being called in an exhibition game in November 2005. "Joe Alleva told me that they are trying to schedule Marathon Oil for a preseason game next year, and we are in discussions about calling another foul against Duke at that time."

Game Notes: Dick Vitale and Brad Nessler will be handling the TV play-by-play for ESPN. Vitale says he has no idea what he will say when the foul is called against Duke but he has been placed on a prescription sedative as a preventive measure.

The score went to 21 to 10 in Duke's favor after Williams nailed another field goal and free throw. He was making mincemeat of the Hokies' inside players, having already hit six of six field goals and scored 14 points.

At 13:29, JJ Redick went to the free-throw line and actually shanked one, only his fifth miss of the year. He was shooting close to 94 percent at the charity stripe. He then dropped the second.

With ten minutes to go in the first half, Virginia Tech had four starters saddled with at least two fouls each. A foul was called on Tech guard Jamon Gordon, then another on Carlos Dixon. By the time Duke had gone up 36 to 25, Tech had already been called for its seventeenth team foul. The Crazies started chanting, "Please stop fouling." Shortly thereafter, Tech earned fouls 18 through 20; Duke had been whistled for just eight. Now the fans were chanting: "It's not football, it's not football."

"I'm not sure I care about watching this game," my mother said. "This slaughter. This rout. This unfairness."

JJ Redick hit a 25-foot three from the top of the key to close out the half, finishing with 15 points. By now, it was 56 to 31. Astonishingly, he'd also missed two more free throws, news in itself. In his 86 games as a Blue Devil, he'd never missed more than one. This was not a problem in the larger scheme, however; Duke had gone to the line 35 times in the first half.

The second half was more of the same. Duke ended up winning, 100 to 65. Redick finished with 29 points, Shelden Williams with 25 and 17 rebounds. Seth Greenberg finished in jail, as it were, ejected from the game after getting called for a second technical with 4:39 in the game. His unhappiness with the officials may have had something to do with the disparity in fouls. His team set a dubious school record with 34, only seven short of the ACC record. The Blue Devils shot 49 free throws, Virginia Tech 23.

In his postgame press conference, Greenberg complained that as he exited the court, he'd just missed being poked in the eye by a Cameron Crazy happily taunting him. "This is a magnificent atmosphere, a terrific building, and obviously something most coaches are envious of, the atmosphere they've created here. But there needs to be some security. When someone's walking off the court and people's hands are in his face, to me it's wrong, and it needs to be addressed. It's a potentially volatile situation."

When asked what he and Greenberg were bickering about at one point in the second half, Mike Krzyzewski deadpanned, "We were talking about expansion."

My mother had disappeared into the kitchen midway through the second half. She couldn't bear to watch anymore. We were rounding toward the first regular-season game between Duke and North Carolina, and our morale was volatile, to say the least.

BUFF ME THE HEAD OF DICK VITALE

In the Cold War bilateralism of the Duke–Carolina rivalry, in which the friend of my enemy was my enemy, ESPN announcer Dick Vitale had earned my enmity and that of Carolina fans for his hyperventilating, on-air slobberfests on behalf of all things Duke—so much so that even Arizona coach Lute Olson had started referring to Vitale as Dookie V. A Carolina student had recently told me about a Dick Vitale drinking contest played during games on ESPN. Every time Vitale mentioned Duke in any capacity, the students had to drink a beer. The wrinkle was that the drinking game was played only during matches in which Duke was *not* playing. "We're usually drunk by halftime," the student told me.

I went up to campus one afternoon toward the end of January to lie in wait for Vitale at The Bull's Head Bookshop. He was in town to sign copies of his book *Living a Dream*, foreword by Mike Krzyzewski. You scratch my bad back, I'll scratch your bald head.

As Vitale barreled into the store, it was as if shock winds from a nuclear explosion had blown open the plate-glass doors, riffling the bluebooks and newspapers and whipping our hair into tangles. He made a beeline for a blond coed and had wrapped her in a bear hug before she even saw him coming. She looked up in astonishment to see that Vitale had ahold of her, then relaxed into his arms with granddaughterly ease. "Got a date tonight?" he asked. "Ah, give me a shot." He called out a guy who seemed to be hovering proprietarily around the blond. "You've got no shot, because she's my date," he said. He pointed to a gangly fellow who'd been watching Vitale. "He's looking at her, everybody, not me." The gangly fellow grinned in pain.

And so Vitale bulled through the bookstore, swooping around in his blue crewneck sweater, a bald-headed unguided missile of omnivorous exuberance, looking for a place to detonate. Or nuzzle. Or be nuzzled. A scrum of frat boys began rubbing Vitale's famously shiny head. "You guys messed up my wig!" he screamed, smoothing his fringe.

"Go call your mothers!" he yelled at the students lining up at the table where Vitale would sign books. "Tell your parents that you love them! Make good choices in your lives! The key to life is making good choices! I'm stealing money, because I get to watch games from half-court! I'm sixty-four-years old, acting like I'm eighteen! I'm a blessed guy! Life has been good to me!"

The great thing about Dick Vitale was that he gave you the Dick Vitale experience as reliably as Kellogg's gave you a bowl of cornflakes. He lived the brand. He met the theatrical standards he had invented for himself on ESPN. Every student, every college basketball fan in the universe could "*Awesome, baby!*" you to death in imitation of Dick. (That's how he was autographing the books, actually: "To ___, You're awesome, baby!") He was everybody's favorite uncle, the one who'd pull quarters out of your ear when you were a kid and slip you a twenty when you were a teenager. Despite his decades as an announcer, he loved basketball like a fan who sat in a scuffed recliner all weekend, gorging on games. College basketball wasn't a profession, it was his vocation.

As with saints and lamas and various heroes of integrity and spirit, there seemed to be no discrepancy between his inner and outer life. He was a highly paid missionary for Vitalism, that philosophy of exuberance, with its three-word mantra: *super, scintillating, sensational!* With Dick Vitale, what you saw was what there was. And that was man as golden retriever, bounding around, licking people, nuzzling, barking, demanding to be looked at and loved and patted on the head. You couldn't really dislike a man who acted like a golden retriever. Or could you?

A student presented him with a copy of the book. "Do you have the audacity to write, 'Duke sucks'?" the kid asked. The dozens of students in line went nuts, roaring their approval.

"Oh, no, I can't do that," Vitale moaned.

"Why not?" the kid asked.

"North Carolina will be back!" he answered. "You better believe it!"

A gray-haired, goateed man crept up to the table. "I'm not kissing you!" Vitale yelled. The man whispered something in Vitale's ear. "Hey, we've got a 'Zona fan here," he blared to the students. The man ducked out of the store with a furtive look on his face.

A girl came up and huddled with Vitale. "*Go Duke*?" Vitale yelled, now catering to the Carolina students. "Hey, this isn't right! I've had three in a row that want me to sign, 'Go Duke!' That's not right! That's not *riiight*!" To a smattering of boos, the girl beat a chagrined retreat.

Then Vitale looked up from a book he was signing and started shouting to the crowd. "Dean Smith used to say to me, 'Why do you say so many good things about us?' You guys don't want to hear me talk about Duke, but I talk about people who are successful." He began to recite Duke's stats. "Five straight ACC titles . . ."

His tirade was interrupted by a candidate for student-body president named Fred or Fawad or something, a little guy in a suit. He wanted Vitale to pose with him in a photograph. He held a crudely lettered sign made out of notebook paper that said *Vote Fred* or *Vote Fawad*. The print was too skeletal to know which.

"I'm voting against you!" Vitale shrieked. "You didn't buy my book!" But he posed for the picture anyway, embracing Fred or Fawad. "What are your issues?" Vitale asked. "Make a speech!" But Fred or Fawad was no Vitale. He couldn't just make a speech. He slunk out of the room, defeated by the hurricane gusts of Vitale's energy.

After an hour or so, the line had dwindled, and Vitale's handler, a well-groomed fellow named Jonathan whom Vitale had been needling for not setting an alarm clock in the morning, passed Vitale off to me. Not knowing how much time I was going to get, I hit him right away with the Duke questions.

"What effect do you think you have on recruiting with all your praise for Duke? You know, don't you, that they call you 'Dookie' V.?" Was I an asshole or what? The sour-spirited, unforgiving partisan had leaked out. The beast was in town, ladies and gentlemen.

"Huh? What? Ah, man, you're kidding me!" he said. "'Dookie V.' Ah, man! You're killing me here. I don't buy that. I don't think I give Duke an advantage. Kids want to play for a winner. Ah, man!" If I wasn't mistaken, this was the first time all afternoon that a little shadow crossed Vitale's face. A slight decrease in his energy field. In spite of the beast, I felt kind of bad.

I also felt kind of bad because having a private conversation in the bookstore with Dick Vitale was no different from listening to him do a game on ESPN. He was broadcasting to me from six inches away. Students on the other side of the store were listening to every word.

"I praise whoever's playing," he shouted. "Come on! I've given all kinds of praise to North Carolina over the years. So I don't think it affects recruiting in any way. The bottom line is whether you're successful. Kids want to be among great players in successful situations. Dean Smith had no problems. Matt Doherty had no problems recruiting."

He lauded the Cameron Crazies. "They're being emulated all over America," he said. "It's really, truly what a college situation is supposed to be. I see teams that come in there that get intimidated once they get down and really feel it's over. Think about the money Duke could make if it put those seats up for sale to the alumni. But they keep the students down there on the floor. That's what makes it so special."

"They did boo Jeff Capel one year," I interjected. I just couldn't decide what kind of journalist I was. The Sam Donaldson of the genteel Bull's Head Bookshop, who, instead of asking important questions about political deceit and national defense policy, was grilling an enthusiastic sweet grandfather of a man about aiding and abetting Duke University? To this my life had come? The manager of the bookstore stood nearby. She looked concerned. I was making the guest of honor unhappy.

Vitale gave me a look. I thought I saw a flash of pique cross his eyes. What kind of interview had Jonathan gotten him into? Who was this sourpuss? "Right. Right. There were boos," Vitale said. "But the point was, it was still spirited."

Then, in true Vitalian fashion, he took it upon himself to dissolve the uncomfortable vibe. His enthusiasm was dictatorial. It brooked no darkness. He decided to tell me a story about—who else?—himself.

"One time, there was a Duke–Carolina game, and I ended up cutting my head," he volunteered. "We're working upstairs at Duke. And it's really tight and warm up there. And there's this metal hanging down, this beam or something. This is the game that went into overtime— 1995! That's when it was! And Jeff Capel hits the long shot. Well, I jump out of my seat and I bang my head and as I was bleeding, I just went on doing the game. Blood dripping down my head. And my broadcasting partner, Mike Patrick, says, 'You never even blinked.' I wished that game had never ended. I wanted to stay all night. My flight's at six the next morning. Let's play *ten* extra overtimes!

"Oh, baby, this is such a great rivalry," he said, drunk again on his own gusto, spiraling into Vitalian rhapsody, poking me in the chest with what may have been affection. Or may not. "I'm a basketball lover, and this is heaven for me. I'll tell you how special it is. I used to take my daughters out of elementary school and bring them here because I wanted them to understand what this environment is all about, how important it is to get good grades so you can be a part of this. It's such a great happening. And it's not just the game. It's all day."

My ears were ringing. I gave up. He had given me his blood. Soon enough, he would be shilling for Carolina, anyway; he didn't just jump on the bandwagon, he drove it. We traded a few more memories back and forth, I pitched him softballs and he knocked them into shallow left field for singles, and then he asked me, "Is that good? Got enough?" And I said, "Great stuff! Thanks, Dickie V."

As I left, he was donning a Carolina blue Afro wig. Students from the campus TV station had asked him to do a promo for them. "Watch this," he said. And he fluently improvised a 15-second spot, finishing as follows: ". . . and the neighbor from down the road will be checking into town!" He meant Duke.

SIX

The Shooter

BEFORE THE DUKE–WAKE FOREST GAME, my mother and I ate supper with the Judge and his wife at the Mexican restaurant at the foot of Strowd Hill. They had been great friends of my parents for decades, and had taken good care of my mother since my father's death.

The Judge asked me, "Will, who do you want to win tonight?"

"Wake," I said.

"I'm hoping there's a way they can both lose," the Judge said Solomonically, with a wry, authoritative look that I was sure must have been very effective during his days on the federal bench.

I had long admired the Judge's capacity for bemoaning the catastrophic state of civilization while simultaneously deriving a good deal of entertainment from the same. This quality was on full display tonight. Shaking his head in bemusement, the Judge said: "Did you see that God had time to help T.O. this week? Between assisting the planets in their course, he apparently decided to intervene in the healing of T.O.'s ankle."

The Judge kept close watch on professional football, and he had been following the saga of Terrell Owens. Owens was the volatile but talented wide receiver for the Philadelphia Eagles who attributed the miraculous recovery of his injured ankle in time for the Super Bowl to God's speedy response to his fervent prayer. "I've got the best doctor of all," T.O. had said, "and that doctor is God."

"I'm going to ask Bob Dunham a question," the Judge said, referring to the preacher at University Presbyterian, where he and his wife were longtime members. "I'm going to say, 'Preacher, how about if you preach a sermon on the power of intercessory prayer and T.O.?'" His eyes crinkled mirthfully. I think the Judge was a skeptic about such prayer. When my father seemed to stir a little in his coma, I remember the Judge having said with the wonderment and regret of a man who actually believed otherwise: "Maybe there is a little something to intercessory prayer."

The Judge and his wife were the kind of Presbyterians I admired. They appeared a tad skeptical about the idea of approaching the Creator too directly or burdening Him with personal requests. If you sprained your ankle, for instance, better to ice it and let God attend to cosmic priorities, like the motion of the planets, the birth of stars, things like that. So many contemporary Christians wanted to call on God as if He were a 24-hour help line.

I was reminded how, many years ago, my father and I were watching North Carolina play South Carolina. As usual, the Gamecocks' John Roche crossed himself on the free-throw line. This never failed to raise his ire. "Do you think God cares who wins this basketball game?" he asked.

I agreed wholeheartedly with my father on this—they offended me, these exotic Catholics from New York asking God to favor them at the charity stripe over us double-predestination Presbyterians, who wouldn't openly presume God's favor and who just sighted our free throws and let them fly. (Whether they went in or out had already been decided.) However, it was this particular son's fate to probe for loopholes in his father's encompassing worldview. "How about if it were the finals of the ACC tournament?" I asked. "Maybe God would care who won that one." I knew that I did.

"Honestly, Will, do you think God cares who wins the ACC final?

Don't you think He has better things to do than worrying about a basketball game?" My father couldn't always tell when I was just playing with him. In fact, I couldn't always tell myself. These things could turn on a dime. So that what sometimes started out as me being facetious ended up with his rebuking me and me getting serious.

On the other hand—there was always an "on the other hand" with me (it must have been exhausting for friends and family)—it was clear to my 11-year-old mind that if God didn't care about who won the basketball game, then who was to say He gave a shit about an individual sparrow falling from the sky (the classic example)? Why would He care about one thing and not another? It seemed that us humans were imposing our sense of moral proportion on Him. How did God decide what was important enough to care about when He was so powerful that nothing escaped His notice? When I was 11 or 12 and missed a couple of jump shots in a row on our hillside basket (and had to chase the ball through the brush to the creek at the bottom), I accused God of wanting me to miss. "You could have made it go in if You'd wanted to," I told Him. This was the sort of cross an all-powerful God had to bear.

I may have inherited my freethinking ways about these matters from the most ardent defender of the church in my life: my father. Theologically, his mind was wonderfully complex and brave. "The problem is that most people can't live with the lack of certainty," he often told me. The implication was that, like it or not—and I don't think he liked it— he *could* live in doubt.

Most churchgoers, at least the thinking-feeling ones, were probably more like my father than not, their religion more idiosyncratic than the dogma they ostensibly believed. They inhabited realms of private prayers and desperate bargains; they were believers in breathtaking heresies that would have gotten them burned at the stake in doctrinally purer times. "I think God is love," my father said to me one summer. "But that is all I think God is. Just love."

Now, in the earthly realm of Joel Coliseum in Winston-Salem, JJ Redick had just driven down the pike for a basket, prompting Dick Vitale to hyperventilate for the fifty-seventh time this season that "you can't just play JJ Redick for his outside shot, you have to respect his willingness to put his head down and drive toward the basket."

"I don't know. Maybe we should switch it to Bush," my mother said. Given my mother's feelings about George Bush, Vitale had driven her to extremes—the idea that it might be better to watch the president deliver the State of the Union address than endure another second of Vitale's State of the Dukies rant.

"Mama, you just have to let it go in one ear and out the other," I said.

Admittedly, this was hard to do. Shelden Williams had been fouled going to the basket, cannily using his back to throw off defenders like an irritable stallion. And yet, Vitale paid no mind. "The key to the Duke season," he was intoning with that Vitalian exuberance that would be used to convey precisely the opposite opinion within a week or two, "is the development of Shavlik Randolph."

Sometimes the Vitalian psycho-speed-shift arrived even faster than that. JJ Redick curled around the lane, bumping his defender off multiple screens, received the ball out high, and took a rushed, off-balance three-pointer. Vitale went ballistic, ripping JJ for bad shot selection. Mike Krzyzewski, however, remained calmly seated. Unlike Vitale, he knew the truth about Redick: For him, there was almost no such thing as a bad shot. Thirty-five feet out, falling out of bounds? Take it. A desperation shot for most shooters was a plausible shot in Redick's hands. Such players extended the boundaries of the reasonable. It was a fact that even Carolina fans had to concede.

The game was sensational. In the second half, with the score 45 to 43 in favor of Wake Forest, Redick hit a three. But on the other end, Wake's Vytas Danelius canned one of his own. With Chris Paul playing brilliantly, weaving through traffic with the ball like a scatback, the Demon Deacons proceeded to go on a run. When Paul knocked down a three to extend the lead to ten, Vitale began his traditional shriek: "Get a time-out, baby, get a time-out!" Krzyzewski obliged.

Vitale and Mike Patrick launched into their usual duet of "What's not to like about Mike Krzyzewski and Duke?" The fact that they felt it necessary to deliver these routine impassioned defenses of the Blue Devils indicated that they were preaching to the *unconverted*. Perversely, this was sweet music to my ears. Not to my mother's. "Do you think Bush is still blathering on?" she asked hopefully.

The Duke time-out didn't help. With 7:46 to go, Wake was still playing well and had gone up 75 to 61 as Paul scored on a layup, twisting and turning on his way to the basket to avoid the charge. During the Deacon surge, Redick had even missed the front end of a one and one.

Wake Forest appeared on the verge of knocking the Devils out. The Vitale Extraneous Jabber Index indicated as much. He was now talking about next week's Duke–North Carolina game. "I believe it's the greatest rivalry in all of sports," he said.

"Even when he says something that makes sense," my mother said, "he says it at the wrong time."

It seemed that Wake had the game safely tucked away, until the last two minutes, when JJ Redick went out of his mind. He nailed three three-pointers and a herky-jerky drive to the rim. After the shot that trimmed the Deacon lead to a mere basket at 86 to 84 with 1:04 remaining, Redick landed on his rear and went sliding down the floor. "One of America's great, great shooters of all time!" screamed Vitale. "He can shoot it from our studio in Bristol, Connecticut." The Wake crowd was in a panic. Redick hit another, slicing the Wake margin to one, at 90 to 89. With 2.2 seconds left, Duke's Lee Melchionni fouled Taron Downey, who bottomed both free throws to give Wake a three-point lead.

Would it be enough? Taking an in-bounds pass from the reserve center Patrick Johnson, Redick launched a leaning three from near midcourt. The crowd held its breath. As did my mother and I. Time expired while the shot was in the air, falling towards the goal. It missed, and the game was over. Wake had triumphed and Duke had lost, but somehow Redick's stomach-churning brilliance made it feel like the other way around. He had finished with 33 points on ten of 22 shooting, five for 13 from three. "I thought it was going in," the Wake Forest head coach Skip Prosser said of Redick's last shot. "I'm not telling anything people don't already know. But he's a phenomenal shooter."

"Whoooo," my mother said. "That was close."

THE SHOOTER

Many nights, JJ Redick would turn out the lights in his room, put on Bruce Springsteen's *Nebraska*, that haunting, autumnal reckoning of loners and losers, and sit there in the dark, listening.

Sometimes, he might take out the little leather notebook his sister had given him one Christmas, the pages dog-eared and thick with writing, and scribble a poem or two. He read his poems aloud to Shav, and he read them to his family. "I got a few tears out of them," he bragged about one holiday recitation. His sisters teared up, he himself teared up. Redick was always saying this about himself: *I teared up. I cried.* At movies, poems, songs, JJ Redick cried as easily as Bill Clinton at a ribbon cutting.

Yes, perhaps the most hated player in college basketball—*Sports Illustrated* had said so—JJ Redick wept a lot. He wept out of deep feeling for the pathos of winners and losers. He wept out of the strangeness and beauty of life. And then he wept some more.

That he was so hated appeared to be an object lesson—courtesy of television—in mass projection, in the incongruity between image and reality, and in the multitudes contained within one human being. And, frankly, in the way one complex-hearted human being could arouse multitudes *outside* of himself to paroxysms of hatred, to vituperation worthy of Roman sports fans at the original Colosseum giving some poor Christian the thumbs-down. Plus it meant that Redick was a really good shooter. He loved sticking long-range daggers into the hearts of opponents and fans, sticking them in and twisting until everybody got real quiet.

So naturally, if the nation detested JJ Redick, you could rest assured that Carolina fans did, too. He had inherited the Christian Laettner Dishonorary Chair of Most Despised and Accursed Blue Devil, a position actually endowed in the late Eighties by Danny Ferry but named after Laettner (Class of '92), whose invaluable contribution to international Duke hatred established a standard that will probably remain unsurpassed in our time. Previous chair holders had also included Bobby Hurley, Chris Collins, and Dahntay Jones.

Had he been a poor shooter, Redick would have been merely an object of mockery. Because he shot extraordinarily well, he was feared to the point of hatred. Everything about him was subject to derision—his acne, his partying (he was known as JJ Redeyes after being busted by campus security in a room cloudy with pot smoke), and his poetry (a sample of which appeared in a *Sports Illustrated* profile and had inspired a raft of parodies on *Inside Carolina*). So intensely hated was Redick that even his greatest attribute as a player, his long-range shooting, came under assault. Critics carped that he didn't shoot as high a percentage from three as Arizona's Salim Stoudamire (true). He didn't even shoot as high a percentage as Rashad McCants (also true). But given Redick's capacity for striking from just about anywhere on the court, this was like complaining that Muhammad Ali's punch wasn't quite as heavy as Mike Tyson's. It was heavy enough. Redick couldn't win for winning.

He had to be disassembled at every level.

But here's the thing (and this really stuck in the collective craw of Carolina fans): Redick didn't seem to mind. He appeared to love the animosity. He looked as if he drank in the crowd's abuse like a high-protein shake and then spat it back at them in the form of three-point shots. Every time he launched one of his long-range rockets, he inspired the greatest intake of breath since the Cuban missile crisis. His mere possession of the ball in shooting position was enough to spook the fans of rival schools.

Like many of the Tar Heel faithful, I watched anxiously as Redick wove his ceaseless figure eights around the court, rubbing defenders off big men down low, darting for the cover of successive screens like a roach desperately seeking a hiding place. Or, to put his scurrying in a more elegant light, he reminded me—antiquarian that I was—of the former Boston Celtic, John Havlicek, who ran tirelessly across the hardwood like an eight-year-old cutting through neighborhood backyards on an endless summer afternoon. Just the other night, when Duke was playing Wake Forest, my mother exclaimed, "Who's got JJ? Who's got JJ?" as Redick slid off a high screen and popped open in the corner, where he waited luxuriantly for a pass. Her use of his first name struck me as an intimate way to mark an antagonist.

When Redick actually missed a three-pointer, as he had about 60 percent of the time this season (which meant that he actually hit an entirely laudable 40 percent of his treys, even with defenses keying on him), the crowd celebrated as if every last one of them had won a free barbecue sandwich—with extra slaw—from Allen & Son.

I wondered what it felt like to be at the laser point of this mass animosity, to be both despised and feared to such a degree. (I did know a little something about being hated, but that was in more intimate circumstances.) What seemed clear to me was that in some quivering, lively, even generous way, Redick channeled all the Duke hatred so that it ran headlong like a river of acid towards his person. Only Lee Melchionni rivaled him as an object of antipathy, but because he wasn't half the player, he received only about half the disdain.

So one February afternoon, the skies over the Triangle disconsolately Baltic, I went to Duke to meet the most hated player in college basketball. The JJ Redick who showed up at Cameron that afternoon didn't seem like the kind of guy capable of inspiring people I knew well (I mean really well) to scream their lungs out at him. He ambled over to me after practice sporting a battered Boston Red Sox cap, brim cocked skyward, dark hair twisting from underneath like campus ivy, a wry smile on his face. He didn't bristle. He didn't beat on his chest. He didn't pump his fist in the air, as he had frequently in Duke's last game.

In fact, he radiated a positively laid-back bonhomie. He sank into a chair, stretched out his legs. Ease and languor. Maybe competition purged Redick of whatever bile and bitterness can accumulate within a 21-year-old All-American and left him in a state of grace.

By way of opening civilities—and because locker-room banter made me feel sprightly—I told Redick that his teammate Sean Dockery had told me that JJ couldn't score on him. Redick laughed, generous, amused. Was he largehearted because he was good, or was he good because he was largehearted?

In fact, Dockery had told me that he and Redick had nearly gotten into fistfights a couple of times in practice. "I'll stick him," Dockery had said, "and he'll say, 'Come on, Sean, you fucking hacked!' And I'll say, 'Don't think you're going to score.' After practice, we laugh about it. But in practice, we probably can't stand each other."

Redick admitted that his freshman year, he and Dockery had nearly come to blows several times. The next year, they still fought each other throughout the Duke practices, holding and grabbing, but when the whistle blew, they tended to pat each other on the butt and say, "Good job, man."

"Yeah," Redick said, smiling. "I tell people they can't guard me." He liked to goad his defender. Most of the players talked a little smack. "We just don't always say it so Coach can hear."

Of course, Krzyzewski was hardly the type to censor himself; his smack could knock a player down. Redick knew that. He had been both the beneficiary of the coach's praise and the target of his wrath.

The February of his freshman year, the team was watching tape of their home game against North Carolina, a contest Duke had won 83 to 74. During the game, a play had been called for Redick. "I made a read off of it," he said. "And it was a good read, but I ended up making a bone-headed move. And I remember Coach looking at that moment on the videotape and looking at me—this is in front of the team—and saying, 'JJ, I love your heart. I love your courage. And I'm going to love coaching you and winning a lot of games over the next four years.' I got teary-eyed when he said it and he was teary himself."

Of course, what the good Krzyzewski gave, the bad Krzyzewski could take away. He possessed the tongue of a drill sergeant. And Redick, praiseworthy on some nights, had felt the lash of that tongue on others. The worst thing that Krzyzewski had ever said to him he was reluctant to even to describe. He was remembering a game during his freshman year when Duke was playing Georgetown. At the half, Redick had racked up precisely zero points. "I was playing just horrible," he said. "Making stupid plays. I went into the locker room and Coach said, 'JJ, you're playing like a girl.' Only he didn't put it quite that way. Substitute something more explicit for 'girl.'"

Hmmm. "Pussy," I surmised. Coach K had called J.J. a pussy.
Redick smiled conspiratorially.

Redick's path to Duke had not been the customary route of basketball prodigies. When he was born, in 1984, the family lived on a communal farm in Tennessee. Not long after, they had moved into a gallery in Cookeville, Tennessee. His parents had both been potters for a time, hence his middle name, Clay. "They weren't hippies or anything," Redick said, at pains to dispel an impression I wouldn't have been bothered by in the first place. Every family has its own myths. One of the Redicks' foundational legends involved "the hammer." It was the mid-Eighties. JJ was two. His father, Ken, gave him a hammer.

Young JJ loved that hammer. And one day, when his father was outside mowing the lawn and his mother was upstairs, JJ took his hammer and proceeded to smash to bits the tile table his father had built. "Because that's what hammers do, man," Redick said with a laugh nearly 20 years later. "They smash."

These days, he said, his father claimed that he had given JJ the hammer for hand-eye coordination. Why not? The boy must have gotten his shooting eye from somewhere.

From Tennessee, the Redicks moved to Virginia, where they now resided in Cave Spring, just outside of the city of Roanoke in the western part of the state, where Ken made a living in health care and his mother, Jeanie, worked as a nutritionist. He inhabited a mostly white world there. His senior year at Cave Springs, the varsity he led was all white. Leon Curry, whose son JamesOn had signed with North Carolina before losing his scholarship on drug charges, had been impressed by Redick's fearlessness. If white players sometimes had a reputation for being soft, that hadn't been the case with Redick, Curry said. "He carried that high school team of his against Danville George Washington. Those boys didn't scare him. He dropped thirty-five points."

Outside of Cave Spring, the racial mix of Redick's world was reversed. He was often the only white player on Boo Williams's All-Stars, a top-flight AAU team based in the Virginia Tidewater. His teammates, he said, "were amazing guys, and they really embraced me."

Redick regarded himself as deeply at home in the hip-hop culture that pervaded American basketball at all levels. "Growing up," he said, "I probably wasn't really comfortable with it at first. I remember when I was eleven and I was playing with a city team in AAU ball and we were driving back from a tournament. Someone, I can't remember who, said, 'Hey, listen to this.' And he put some music on and it was Warren G. It was the first time I had ever heard rap music and so I'm listening to it and I'm hearing the F-word and I'm like, *Wow.*"

Now he numbered among his heroes three rappers, Nas, Eminem, and, especially, Tupac Shakur, whom he esteemed most highly of all and who (along with Springsteen) influenced his own writing.

There was a swagger to Redick's game that seemed more street than suburban, full of hip-hop bravado and glee, and he was pleased that people thought that on the court he exhibited a kind of black style. "I've had people tell me that," he said. He laughed. "I mean, if you see what I wear off the court, you'd be like, He's a white guy. Jeez, look at me: I've got a choker necklace and a faded cap."

"That's pretty white," I said.

"So frattish," he said, fingering his choker, mocking himself.

"So I guess there's two sides to me," he concluded. "I've always felt comfortable around people of other races and backgrounds. I'm a people person. I feel like I can fit in to just about any crowd."

He'd committed to Duke before his junior year at Cave Springs, at the time the earliest pledge Krzyzewski had ever accepted from a recruit. His first love as a kid was actually baseball; Duke basketball was his second. He'd wanted to play for Duke since he was eight. He estimated that he had played in at least 50 Duke–Carolina games in the anonymous glory of the Redick backyard.

I asked him if he always hit the last-second winning shot.

"No," he said with a laugh. "Sometimes it was a blowout."

In a sense, he was just one more boy dreaming of playing for his childhood team. Only in his case, the dream had actually materialized, and he appreciated the good fortune of that. Despite his confidence, he did not seem a player who considered his success preordained. "Just to dream of something that was so amazing and to watch it on TV and to have wanted to be a part of all that—and then for it to become a reality! Man! When I play in a Duke–UNC game, I'm still living out that dream."

Mike Krzyzewski had fallen in love with Redick's play at a summer tournament going into the player's junior year. "I knew he was a special player right away," the coach said. "And I saw people follow him, so I thought he had leadership potential. I thought he would learn from Mike Dunleavy, but then Dunleavy left."

Now that he'd had Redick on campus for three years, Krzyzewski had come to see him more precisely. He wasn't so much the leader that the coach had initially expected. He'd been caught by campus security in a dorm room in which students were smoking pot. Redick himself acknowledged that in his first two years, he'd partied more than was good for him. However, in the past summer, he had appeared much more focused. In summer school, he'd gotten obsessed by the Civil War, courtesy of a class given by the Duke professor Jim Barker, whom Redick called the best teacher he'd ever had. The class toured a couple of battlefields in Virginia. "I actually stole a Confederate flag from Jeb Stuart's grave at Hollywood Cemetery in Richmond," Redick confessed. "I feel kind of bad about hanging it in my room, though." He'd also dedicated himself anew to the game, paring away distractions, narrowing his social circle, cutting out late-night junk-food binges. A program of intensive running had reduced his weight from 215 pounds to 190.

"He's learning," Krzyzewski said of Redick's commitment to being a team leader. But the coach didn't want his charge's expanding role to take away from his primary responsibility as a shooter. "If you're a scorer-shooter," Krzyzewski said, "you have to be unbelievably focused. Shooting is an art. And you have to be a little bit selfish, you have to be a little more self-centered and more focused about that. So I want JJ to lead without taking away from what he does. I mean, he could become this leader and lose his focus as a shooter. There's a balance that has to be maintained there."

In July 2004, he'd gone out to California to work for ten days as a counselor at Michael Jordan's camp. At night, the counselors—all top-flight players—would put on exhibitions for the campers, "kind of a show," Redick said, "guys throwing alley-oops, blah, blah, blah." Then the campers would leave. That's when the games got fierce. And that's when, at one point, Redick found himself guarding—and being guarded by—Michael Jordan, "my favorite player," Redick said, "the favorite basketball player of my generation."

Jordan had put on a little weight, Redick said, "but he was still easily the best player on the floor. He was killing, and letting people know about it, too." Naturally, the former Tar Heel teased Redick for being a Blue Devil.

On one possession, Redick switched off and found himself face-to-face with Jordan. Or, more accurately, face-to-backside. "He posted me up," Redick said. "He gave me the little head fake, a little side step, and then shot a turnaround fade-away. As soon as he shot it, I was like, 'It's off! It's off!' He missed it."

Another time, Redick nailed two deep threes in a row over Jordan to tie the game. "As I was going down the floor, I kind of stared Michael down," Redick said. With a grin, he added, "I don't think he saw me. And if he did, it didn't faze him, because he hit the next two shots to win the game."

What caught Mike Krzyzewski's eye even more than players following Redick around a summer camp was JJ's shot. His threes arced towards the basket as if possessed of memory, as if, like a flock of arctic terns, they were simply following their traditional migratory route home. Redick flicked the ball basketward. The ball soared and fell, spinning in reverse. And even after it descended through the net with the faintest whisper, and Redick was already backing down the court, he continued to hold his wrist aloft in an exaggerated follow-through, inscribing in the air the silhouette of an Egyptian dancer on a tomb wall. Such strut— it drove many fans and players crazy. They thought Redick was rubbing it in with that follow-through.

The maintenance of such a shot called for regular habits. If he had a nine o'clock game at home, for instance, Redick's schedule was as structured (if a tad more leisurely) as the president's. During the day, he went to class. If he didn't have a class, he might take a short nap, although he didn't really like taking naps on game days. The team shared a meal about four hours before tip-off. After that, Redick returned to his room, put on some music, and relaxed. About two and a half hours before the game, he left his room to find a cup of coffee. He did this whether the game was home or away. At Duke, all he had to do was amble across the street from his dorm to the Starbucks. Then it was time to head to Cameron, where he'd sit in the sauna for a while, then take a hot shower. The heat loosened him up.

Once he reached the court, his habits were similarly unvarying. Like most great shooters (and many mediocre ones), he was as faithful to his routines as an obsessive hand washer. It wasn't just that practice made perfect—practice entitled a player to approach the altar where good fortune was handed out. Redick attributed his success as a shooter to three things.

The first two were form and repetition. "I think you've got to have good form," Redick said. "But I've seen guys who don't. Michael Redd [of the Milwaukee Bucks] doesn't have good form, but he's a heck of a shooter. And then repetition—you should just catch the ball and shoot it the same way every time. You don't want to be thinking about it." Like many great players, he intended to bury self-consciousness in a deep grave from which it couldn't dig its way out during a game. Thinking was death itself.

Above all else, Redick believed in confidence. And this was where the material world affixed itself to the spiritual. The worst shooting slump he'd experienced was from the beginning of the 2004 season until the Texas game on December 20, 2003, in which he'd scored 20, going five for nine from the field. Wincingly, he recalled that during that period, he'd shot only 29 percent from outside the arc. He'd missed three months of basketball from July to October and was having to fire his way back into form. And yet, while it had irritated him to have people coming up, asking what was wrong with his shot, he had never lost confidence. "I always knew I'd break out of it," he said.

And he always had. The capricious gods of shooting never forgot him.

Great shooters believed in voodoo, in the preeminence of feel. The mastery of technique just got them to the beginning. The end was reached when they could launch a shot in the dark, knowing that it would ripple the nets like an evening breeze. Repetition allowed a shooter to forget what he was doing. Feet balanced, knees bent, strong jump, elbows in, wrist flicked, index finger pointing towards the basket—for Redick, these were only means to an end, the physical offerings to the gods of the three and the free.

Before each game, Redick started shooting underneath the basket, banking shots two feet away from both sides, just to get his form down. Then he moved out a little farther, making a semicircle of shots from around the lane, watching the ball go in. After a time, he worked his way out to ten or 15 feet from the basket, popping jumpers. When he became comfortable in each spot, he moved back. Eventually, he began taking threes.

The most threes Redick had ever nailed in a row just shooting around was 47. When he undertook drills in which he took five threes from five spots on the court, it wasn't uncommon for him to hit all 25.

Despite his lack of quickness, Redick clearly presented considerable difficulties to defenders. His brilliance as a jump shooter set up the rest of his game. When his man guarded him tightly, which made sense—why let him get off a jumper?—Redick took him to the basket, something he had been doing with increasing frequency this season. If you played him too closely, you risked fouling him. And it was a foregone conclusion that Redick, on pace to set the all-time NCAA record for percentage from the line, would punish you with free throws.

His forays to the basket weren't especially pretty; he usually looked like a fullback plunging into the line, shouldering into the defender, and pushing his way slowly forward. You could go fix a snack in the kitchen and still have time to catch the end of the drive. But given that Redick often drew double-teams at such moments, he had also expanded his game by becoming a dangerous passer, feeding teammates who'd run to the space opened up by his maneuvers. Daniel Ewing and Lee Melchionni were usually the prime beneficiaries.

Redick believed that the ideal way to defend him was to face-guard him the entire game. But if a team did that, "somebody else is going to kill you," he said. "We've got so many other guys that can shoot."

In the games against the Tar Heels last season, he'd been guarded partly by Melvin Scott, one shooting guard against another. "He's a very solid defender, very alert guy," Redick said. "He's six-two, though." His nemesis on the Tar Heels was Jackie Manuel, the fluid, six-five Floridian. "He's the toughest defender I've faced in college," Redick said. "He's just so long, and he covers my shots so quickly."

In the first game at Chapel Hill in 2004, Manuel had blocked two of Redick's shots in the first half. The whole second half, he said, those rejections "lingered in the back of my mind." As a result, he rushed his shot, racing around screens and hoisting threes before he was ready, the elongated shadow of Jackie Manuel inseparable from his own.

So congenial was Redick on this February afternoon, now fading into cold dusk, so stretched out and at ease, that it was easy to ask the question that I'd had on my mind.

"So, how does it feel to be the most hated man in college basketball?"

He laughed. He regarded hatred as psychic fuel. If people hated him, he must be doing something right. "I enjoy people yelling at me," he said. "I feed off it. As negative as it is, it gives me energy. I really enjoy away games. It's kind of like, hey, I'm going against this team but I'm also going against fifteen thousand people that want to see me fail."

Last February in Chapel Hill, a student kept yelling at Redick that he was "the poor man's Matt Lotich," the Stanford guard also known for his jump shot. "That was so random," Redick said. "That was pretty funny to me."

Less amusing were the fans that screamed invective about his family, "really disgusting stuff," Redick said. "Like claiming they've been with my sister." He rarely responded to such taunts. "What can I say to that?" he asked. "I mean, yeah, I really bet you have."

He acknowledged that Duke was the school that people loved to hate. I had never met a Duke coach, player, or graduate who denied that the hate existed. "We're perceived as the place where we live in our own little world," he said. "Private school. Gothic wonderland. Krzyzewskiville. Coach has this big tower. Cameron is this small, exclusive gym. People think we get all the calls. And with the success we've had, it just breeds jealousy. I think that's where the hatred comes from."

SEVEN

Frogs in the Swamp

BLUE DEVILED

Cameron Indoor Stadium

February 7, 2005

3:45 P.M.

EVER SINCE I HAD COME BACK to North Carolina, I had tried to get to Mike Krzyzewski. There were a few questions I needed to ask him, but mainly I just wanted to see what it felt like to be near him, whether what was left of my hair stood up, or my skin prickled, or I got the shakes or the cold sweats or began to break out in hives the way folks are said to do in the presence of evil. With an agenda like that, it was no wonder that the Duke press office, headed by the wry and watchful Jon Jackson, kept heading me off at the pass.

On the week of the first regular-season game between Duke and North Carolina, however, Jackson offered me a sop. He invited me—the me masquerading as a journalist—to Krzyzewski's press conference. K doled out audiences to the media about as frequently as George W. Bush. It wasn't that he was bad with journalists. He just didn't particularly like them or need them. But with the game this week, a collision with reporters was unavoidable.

So on Tuesday afternoon, I slunk into the media room at Cameron. It was the beast in me who made me feel I had to skulk around; like a secret agent, he used the journalist for cover, and there was no telling what he might do if given free rein. Would the polite Southern reporter prevail in their power struggle, or would the beast stand up and make an ass of himself in public the way he did in front of the TV set at home? From the beast's point of view, that Southern-gentility business was a load of horseshit. Why be a journalist if you couldn't be an asshole in public? To get at the truth, you sometimes needed a crowbar.

Maybe I seemed a little frazzled in the pressroom that day. My divided soul gave me no peace. I started eavesdropping on an exchange between two local reporters, Bill Brill, the past president of the U.S. Basketball Writers Association, who enjoyed close ties to the Duke program, and Al Featherston, who until recently had been a longtime writer for the Durham *Herald-Sun* and was now freelancing for the *Duke Basketball Report*.

They were discussing the abiding issue of whether Duke benefited from an unfair advantage at the free-throw line, which was really an argument about whether referees unfairly favored the Blue Devils, which was really a debate about whether Mike Krzyzewski tongue-whipped officials until they did the "right thing." Featherston was working on an article suggesting that winning teams in the ACC tended to be awarded more free throws. "In 1998, Carolina shot 187 more free throws than their opponents," Featherston said. "That was the highest differential within the ACC over the last ten years."

They were interrupted by a television reporter loudly proclaiming to someone: "You're talking about golf lessons? What a waste. Save your money for strip joints."

Suddenly the talk stopped cold. A chill wind blew through the assembly. The room lights blinked. A TV guy's camera momentarily stopped working. "I can't get an image," he said, bewildered. The dark prince had arrived. As the ancient legends reveal, such beings are able to assume appropriate human form. Today, he was wearing a blue warm-up suit with white piping. His cheeks were pink. And as was often said about dictators, rogues, and scary satanic types, he was really quite charming.

First of all, he complimented the opposition, showing both magnanimity and realism. "I don't think anybody gets it up the court as fast as North Carolina does," Mike Krzyzewski said, sipping from a can of Diet Coke. "We can't practice for that. If we could, we'd have Raymond Felton on our second team." He grinned. "And we'd be pretty good if Felton was on our second team."

The press ate this up.

"You have to play against this kind of speed," Krzyzewski said. He was reminded of how the Blue Devils used to take on national teams from Russia during the preseason because their transition offenses were so great. "Our guys would think we were practicing well, and then, boom, in the game, there'd be four to five layups scored against us."

The main thing for Thursday's game, he emphasized, was for Duke to take good shots and not to turn the ball over. "With bad shots and turnovers," he said, "you have no period of adjustment." His current team was good, he said, but not good enough to safely lose its focus on the next assignment. At the same time, he didn't want Duke to become too caught up in the North Carolina game. "I've asked a lot of this team," he said. "We're not deep, with just Lee Melchionni and DeMarcus Nelson coming off the bench. I don't want to wear them out emotionally as we prepare for the off-season. Er, the postseason."

He also had to protect not just the team's morale but the constitutions of individual players. He was pleased with JJ Redick's recent play. "He's taken his game to a higher level," Krzyzewski said. "I have to make sure that our practices don't drain him. He's in great shape, but there's a lot of contact, a lot of dings." As for the benighted Shavlik Randolph, the coach accentuated the positive. "He's trying to do what he can do. Mentally, he's the best he's ever been. He's still feeling the lingering effects of mono."

A reporter asked Krzyzewski how his defensive tactics had evolved since his days as the head coach at Army. "We have better players," he said to laughter.

Oh, he was good. Just modest and playful enough to seem like one of the guys, full of banter and wisdom. Where had all the vitriol gone? "I'm more mellow," he said. He was smiling as if to say, *Me, mellow? Yeah, I know. That wouldn't be the first word that came to mind. But compared to the Eighties . . .* "I understand things better," Krzyzewski said. "When you're younger, you can be more demonstrative about your passion to win. It comes out of more pores." When a writer's cell phone rang and he rushed to silence it, K joked (as if to prove the point), "You can answer it. I'm okay." Like a point guard illicitly cradling the ball, Krzyzewski had everyone in the palm of his hand.

And he could feel it. So he began to wax philosophic. He was an elder statesman of the game now. And he could look at the Duke–North Carolina rivalry with geo-global distance. "In the late Seventies, I didn't know exactly where Duke was. I knew it was in North Carolina. I had a perception of the ACC, but not of the rivalries," Krzyzewski said. "When I first got here, it was like being behind enemy lines. Duke had a better image nationally than we did locally."

"I really believe that this is the best rivalry in sports," he said. "This transcends individual coaches and players. It wasn't about me versus Dean or Dean versus Vic Bubas, and it isn't about me versus Roy. The rivalry is here whether we're here or not, and we're lucky to be a part of it. When you put these two names together—Duke, North Carolina—each individual name prospers. They no longer talk about the Lakers and the Celtics. But this very week, people from different regions of the country will say, we're gonna watch a Duke–Carolina game. They'll tune in, all the Duke fans and all those people who don't like us and want us to lose. This will be here forever."

History, prophecy, perspective. A wide-angled, long-term view of things. Winning and losing were just part of a larger whole. Without a great rival, you could not be great yourself. Was this Phil Jackson wearing a Mike Krzyzewski body suit? The yin, the yang, the Devils, and the Heels. As William Blake wrote—or was it Marty Blake?—"in opposition, true friendship."

Did this largesse of spirit have anything to do with Krzyzewski's recent scare during the Georgia Tech game, when he fainted during the game, collapsing to the floor in stages like a man kneeling to pray? Right on cue, a reporter asked him about his health. "I got checked," he said. "And am being checked. My daughter said to me, 'Dad, basketball is not that important. Get more sleep.' That's where the argument begins. Because it is that important."

As the press conference was winding sweetly down—daughters and Dean Smith, laughter and diplomacy—the beast nudged the journalist and said, "All this elder-statesman rot is driving me crazy. Why don't you ask him about the time he told Dean Smith to shut the fuck up back in 1989? You know, stir this joint up a little bit? Or bring up when he complained about the double standard in ACC officiating? There's a double standard these days, all right. And *he's* the beneficiary of it. Why don't you ask him about *that*? It's your job!" The beast was leering at me in a way I didn't much care for. He bordered on insolent. And yet, I knew how he felt. I just didn't know if this was the time or place to do what he suggested.

"You ask him," the journalist said. "No. On second thought, don't do that."

"You are such a pussy," the beast said.

"No, I'm not. I'm just trying to be polite. I'm not a real journalist anyway. I just play one on TV."

"Please don't try to be funny," the beast said. "It doesn't suit you."

FROGS IN THE SWAMP

The night before the Duke game was unusually warm for early February. As I left the house to go meet up with Melvin, I could hear the frogs in the swamp behind the Essers' former place having a loud old time of it, basking like spa-goers in the syrupy goodness of the muck. It was a false turn in the season in which they were exulting, but they didn't know that. The mud just felt good.

And as pretty much everything did, this struck me as an omen in regard to basketball. *It was too damned early to start singing.* Yes, it had been a glorious season so far for the North Carolina basketball team; an ancient poster on *Inside Carolina*, even older than myself (he'd followed the team since 1957), had recently proposed the 2005 Heels as the greatest UNC team ever. But that kind of talk put us in the same position as those rapturous frogs in the prematurely thawed muck. All of us needed to shut up and hunker down deeper in the mud if we didn't want to die in a cold snap. Frogs and fans alike, we had a long ways to go before spring arrived.

Winter would be back. And we had Duke tomorrow.

MELVIN SHOOTS ALONE

For Melvin, the day before the first showdown with Duke started with a loss—in badminton. On track for graduation in May, he was taking two classes that semester: probability and a required gym class of his choice. He had chosen badminton. He liked racket sports, having played tennis in high school. That morning at Woollen Gym, some guy in his badminton class had whipped him in the first game of a best-of-three match. And in so doing, he'd made the mistake of trash talking. He told Melvin how he was beating him. He mocked Melvin. He rubbed it in. He was simply too delighted with himself. It wasn't every day that your average student got to beat a Carolina basketball player. Even if the sport at which he was beating that Carolina basketball player involved shuttlecocks and flouncy rackets.

Melvin got mad. He removed his watch. He stripped off his wife-beater and played the next game bare-chested, muscles and tattoos on full display. He announced to his rival, "I'm not going to get dirty, but I'm going to sweat a little bit." The result: a beat-down. The formerly smack-talking student scored no points in the second game; in the third and deciding contest, he managed only one.

It was Melvin's turn to do the talking. "I'll beat you and your father," he told his opponent. The student said, "Oh, you're talking trash now?"

Then Melvin turned to the class. "Is there anybody in here who can play me?" he asked. The other students had finished their own matches and were laughing as they observed the rout. When the class ended and everybody started to leave, Melvin said, "Just one more game." And he beat his poor rival again one more time for good measure while everybody stayed to see. Then he said, "I'll see you Thursday."

On the way into the locker room that night, Melvin proudly showed me his name on the awards board. "I got it for defense against Florida State," he said. At his locker, Melvin changed into blue shorts, Nike Retros, and yet another wife-beater.

In the corridor leading to the court, he switched on the game lights. As the bulbs above slowly winked on, Melvin was already shooting. He started, as usual, with layups from two feet away. He took three shots from each side of the basket, then three from the middle of the lane. The ball was not allowed to touch iron. The shots had to be "all nets," as he put it.

Then five jumpers from the free-throw line, five midrange jumpers from each side of the lane. Then five threes each from different spots on the floor. "Come on, girl," he said to the ball.

Then he headed to the free-throw line. He located the nail at the center of the line and lined up his shot. As part of his routine on the nights before games, Melvin made himself hit ten straight free throws. Tonight, he was having trouble. He got up to eight and missed. Nine and missed. Every time he missed, he required himself to start over. He nailed four in a row, then missed. Five. Missed. I worried that I might be making him self-conscious. But of course, if one sympathetic soul was making him nervous, what would it be like tomorrow, shooting in front of ten thousand enemies who would mock him and deride him and curse him and want nothing more than for him to miss?

In the vastness of the Smith Center, this moment was strangely intimate, as if we were two solitary hikers who'd come across each other in the loneliness of some desert canyon. I saw that for Melvin, drills were everything, a mainstay against randomness and disorder. They had to be fulfilled to the letter at all cost. He was happiest working on his game. For many years, basketball had been the organizing principle of his life. I thought of him in his early-morning workouts back at Southern High School, in love with the idea that he was the only guy in Baltimore awake and working at basketball at that hour.

Here he dribbled and shot free throws, again and again. I was afraid my eagerness for him to hit ten straight was somehow audible. The only other sounds were the ball smacking on the floor and, ever so slightly, Melvin breathing.

At last, he landed ten in a row, and I breathed a sigh of relief. But he didn't like the way the tenth shot rattled in. "I have to hit one more," he said.

He missed.

And started again. From the first free throw.

As the ball spun out like a golf ball shirking the cup, Melvin yelled at himself: "What would Smitty say?" Asking what his former coach would say was Melvin's version of "What would Jesus do?"

"What *would* Smitty say?" I asked.

"He'd say, '*Terrible*! Follow through! Don't let your guide hand drift.'"

And as soon as he started channeling Smitty, his form changed. He pounded the ball, sighted the shot, and rose just a little higher on his tiptoes during his release. He held the follow-through just as Smitty had counseled, until the ball dropped through the net and struck the floor. And the ball responded differently, like a creature newly alert to participating in an arc of return, an act of communication between animate and inanimate, hand and leather, leather and net. Now the ball arced cleanly into the net again and again. The net was whipped by the ball's slight wind as it dropped through. *Whffffftttt*. Melvin felt the difference in his shooting, and his muscles locked in on the feeling. Smitty had told him to put such moments into his "muscle memory bank."

Now as he shot the seventh, eighth, and ninth free throws, the ball did the same thing every time. The tenth shot rose into the air and fell through the basket in exactly the same way.

"One more," Melvin said. I might ordinarily have felt he was tempting fate, that he was engaging in an act of hubris against the fickle basketball gods that had finally let him have ten shots in a row. But that wasn't the way Melvin looked at it. He'd gone a long ways by overpowering fate. He'd do it again, as often as necessary. He shot the eleventh free throw, and I needn't tell you that the ball went in. But it did.

YOU CAN SEE THINGS COMING

If Melvin often played the part of a jokester, bantering and busting on his teammates with a swagger he rarely got to display on the court, Jawad Williams, his roommate this season at Summit Hills Apartments, affected a shy, cool distance—laid-back drummer to Melvin's talk-show host—that he attributed to growing up hard in the St. Clair section of East Cleveland. The code there when a guy walked the streets was to hold back, to survey and appraise. When Williams arrived in Chapel Hill, back in 2001, he was thrown for a loop by the friendliness of the students and the townspeople, folks he didn't even know smiling at him and saying hello. "People thought I was unfriendly," he said. "But that kind of thing never happened in my neighborhood."

Their urban roots bonded Scott and Williams. Melvin once took Jawad with him when he went back for a visit to Baltimore. They'd played in a pickup game together until Jawad retired from the fray. "Those Baltimore boys are serious," he'd told Melvin.

At a listed six-nine with the ability to shoot from deep, Williams had been the most heralded of Matt Doherty's first class of recruits, committing in 2000 with Melvin and Jackie Manuel. As soon as North Carolina came calling, he jilted longtime suitor Maryland, who had thought they had Williams locked up. After his brilliant performance at a holiday tournament in Delaware, there was even worry that the forward might leave for the NBA before entering college. In an omen of the battering he would endure in his career at UNC, Williams suffered a broken nose at the McDonald's All-American game in March 2001, which just happened to have been played at Duke's Cameron Indoor Stadium. Figuratively and literally, plenty more blood would soon flow.

Not that the sight of blood should have fazed Jawad Williams. His father and brother were both boxers, both winners of the Golden Gloves. The family was friends with Mike Tyson; Williams remembered visiting his home in Southington, Ohio, where the heavyweight kept a tiger in a cage. Chubby as a youngster, Williams began to train as a fighter. But, as he said: "The first time I got hit was the last time." He was 11 or 12 then, sparring in the ring with his older brother. "I got a few hits off and he told me he was getting ready to pop me. I dodged a couple. The third time, it was over. He got me right in the nose."

Williams began to sneeze. He ripped off the gloves and swung down off the canvas. A basketball court stood next to the ring. His nose still stinging, Williams began shooting. He never boxed again. The bouts had taught him toughness, he said. At Carolina, his psyche appeared composed of equal parts stoicism and vulnerability. He rarely if ever complained, even when trapped in the sustained slumps he seemed to endure once a season, and which seemed to arise from a mysterious swerve in confidence. He usually began the year like gangbusters, nailing threes, finishing with authority on the break. Then, strangely, the former fighter would get popped in the face. In the last days of 2003, he'd been KO'ed playing against Georgia Tech. A long slump ensued. The nature of the injuries he suffered—"They came out of nowhere," he said—made him long for the clarity of boxing, "where you see things coming so you can avoid them."

As a Cleveland native, Williams had never been emotionally invested in the Carolina–Duke rivalry before arriving in Chapel Hill. Now, however, he loved the battles between the teams. He actually listened to the catcalls and the insults raining out of the stands at Cameron in a way he didn't at other gyms. "Last season, after I had to wear the mask," he said, "I heard people there telling me I looked better with the mask on." He noticed a fan wearing a replica of his face guard. He regarded it as a tribute. Better to be noticed and insulted than not noticed at all.

So far, Williams had enjoyed only a single victory over the Blue Devils, in March 2003, in what turned out to be Matt Doherty's last regular-season game coaching the Tar Heels. Most of the games had not provided him with shining moments. In a game against Duke in 2004, he'd missed a box-out in the first half, and the man he was supposed to be blocking out secured the rebound. Roy Williams took him out of the game, but before the coach could say a word to him, Jawad had walked by him on the way to a seat on the bench and said, "My bad."

"I beat myself up more than anybody else," he said.

In the same game, he'd tried to block Chris Duhon's driving layup with six seconds left in the contest. "I had a great angle on the block," he said. "But he came around the side for a reverse."

As always on game day, he'd already put in a heroic effort towards a Carolina victory, for no obsessive-compulsive hand washer was as vigilant as Jawad Williams when it came to placating the fickle gods of chance. He wouldn't forgive himself, he said, if the Tar Heels lost a game because he'd abdicated his pregame rituals, which began in the morning and lasted right up to the tap. When he walked out onto the court, for instance, and he stepped on a line with his right foot—free-throw line, sideline, midcourt line, *any* line—he required himself to step on the same line with his left foot. "That's just one of my rituals," Williams said, "but that tells you what the rest of the day is like."

That evening, he must have missed a step.

THE FIRST GAME
Duke versus UNC
Cameron Indoor Stadium
February 9, 2005
9:00 P.M.

In the media room, a Duke press guy was talking to Steve Kirschner, the affable North Carolina sports information director. They had to iron out an important detail for Roy Williams's postgame press conference.

"Is he a water guy or a soda guy?" the Duke guy asked.

"He's a water guy or a Diet Sprite guy," Kirschner said. "So I guess he's a water guy." They both laughed.

Out on the court an hour or so before the game, Rashad McCants was visiting with his father, who was decked out in McCants #32 jersey and a Carolina-blue hat. Rashad's father talked to him intently. They shook hands for a long time.

One tent after another, the Crazies had been allowed into Cameron an hour and a half before game time. Now they chanted the names of the Duke players. "DEMARCUS [clap-clap] NELSON [clap-clap]." Nelson, a six-three freshman from California, was missing threes even in warm-ups. He'd committed to Duke as a high school sophomore, having been told by Krzyzewski during a visit to campus that he had "angelic eyes," a comment much bandied about for its entertainment value on the *Inside Carolina* website. Nelson rebounded exceptionally well for a player his size, and slashed hard to the basket. But a shooter he wasn't. He appeared fated to become the basketball equivalent of a utility player in baseball, able to do everything well, nothing spectacularly.

Nearby, JJ Redick clanked a three while being lightly guarded. He winced. A guy capable of nailing 47 threes in a row took exception to missing even one. Then, predictably, he proceeded to hit everything he hoisted, fed by a Duke player whose sole job was to feed JJ the ball. As coaches are fond of saying, every player has an important role to play within the "team concept."

Redick finished his warm-ups and ran to the locker room. A high-pitched squeal ensued: "JJ!"

McCants alone remained on the court after all the players on both sides had retreated to their locker rooms. He shot threes. His form was pure. He seemed to be soaking up all the animosity focused on him as he stood alone on Krzyzewski Court. He appeared enlarged by the hatred. He sailed to the basket for a ripping dunk that left the basket shaking. And then he was gone. Defiance. Swagger. The students booed. I was already thrilled. How could the game match such theater? Let's go home. Oh, on second thought, let's wait a second and see what happens.

Now, with the tip-off about an hour away, the joint began to rock with the mass chant: "GO TO HELL, CAROLINA, GO TO HELL." In long rows of bleachers on one side of the court, the Crazies stood. Or, rather, they bounced—bounced like a whole nation of fans trying to hold it in.

The sight lines throughout Cameron were clean, no video boards to tart up the atmosphere. Completed in 1940 for $400,000 and originally called Duke Indoor Stadium, the joint had been renamed for former Duke coach and athletic director Eddie Cameron in 1972, on the day that Bucky Waters's Blue Devils upset a highly ranked Tar Heel team on Robbie West's shot with seconds to go.

There was something old-school about Cameron Indoor Stadium. It summoned up winter nights of basketball in the Forties, long before the days of sneaker fetish and droopy shorts, long before every junior high bench player could execute the behind-the-back dribble, the spin move, the crossover, and the jank in your mug. It evoked Friday evenings at the local gym, farm boys shoveling snow to practice free throws at a basket hanging from a weathered barn, an entire nation of knobby-kneed jump shooters with cowlicked hair, apple pies cooling on windowsills, milk in the bottle, cream skimmed off the top. That's what the architecture, the simplicity, the midcentury economy of Cameron Indoor Stadium, with its 9,314 seats (more seats than Duke students), was saying to me. It was a tender moment of nonpartisan reverie.

Then Speedo Guy brushed by me on his way onto the floor. Speedo Guy was not old-school. Speedo Guy was not apple pies cooling on the windowsill, milk in the bottle, cream skimmed off the top. Speedo Guy was very hairy. Speedo Guy was a little too doughy for public display. Speedo Guy was wearing a tiny brown (not your color, Speed) bathing suit, sneakers, a fake nose, and a blue bra painted on his regrettably pendulous breasts. As a descriptive term, "pendulous breasts" should exist only within skin mags, not within the realm of sporting literature. The phrase "pendulous breasts" should never be necessary in describing a fan. I didn't know whether tonight's Speedo Guy was an imitation or the original, who had first popped up a season or two ago, attempting to distract opposing free-throw shooters, much to the bemusement of the Carolina players, who tended to enjoy the carnival atmosphere at Duke. Given the spectacle, it was a moot point.

What Speedo Guy did on Krzyzewski Court, I regret that I must describe. He gyrated. He shimmied. He shook. He raised his hands. He called for applause, maybe for the Blue Devils, maybe for himself. Maybe the two categories were confused in his mind. Speedo Guy jumped up and down and his pendulous, blue-painted breasts jumped up and down, too. He and his pendulous breasts were not in synch. He jumped, and a second or two later his pendulous breasts jumped, which meant that when Speedo Guy was coming down, his breasts were still going up and appeared to be assaulting him. I wish I could tell you otherwise. But someone had to attack Speedo Guy. It might as well have been his pendulous breasts. When Speedo Guy was finished, he ran off the court and stationed himself directly next to me.

I was sitting under what would become the Carolina basket in the second half. So was a dewy clump of Duke cheerleaders. Speedo Guy cozied up to one of them, bare shoulder to bare shoulder. I think he was simply looking for acknowledgment of his contributions to school spirit. He grinned. When the cheerleader saw who it was, she pulled away, horrified. Speedo looked crestfallen. He also looked very cold, if you know what I mean. I wished the cheerleader had befriended Speedo, despite his abominable performance. Clearly, there was a clash of aesthetic sensibility here. The cheerleaders—now they were old-school. Their sense of theater was traditional. Relative to Speedo's banana hammock, so were their outfits. And given their reputation as the ACC's ugliest cheerleaders, they weren't that bad. Although in that respect, our neighbor Ted on Hillcrest Circle had told me a relevant story. He was an Iron Duke, one of the athletic program's boosters, and once had been invited to help select the cheerleading squad. He clashed with one of his fellow judges over one of the gals she was favoring. "I could carve a better-looking girl out of a bar of soap," he told her.

"I think that's really all Matt Doherty was trying to say," I told Ted.

Now the Duke players came running out of the tunnel back onto the court. They looked as if they were about to storm the beach at Normandy. That stare! Their eyes were pinpoints of flame and focus. The crowd erupted. We were all burning now inside a volcano of noise, a molten core of blare. How tempting it would be to go over to the other side, to merge into this joyous animal froth and frenzy. Just capitulate to the crowd's numerical superiority, its collective happiness, its desire to annihilate the enemy—that is, me and my kind and our Tar Heels—representin', as the kids these days said. Every Crazy was merely one screaming cell in a vast roaring organism. *Join us,* the animal was saying. *You, too, can be one of us. It's easy. It's fun. Don't think. Just do it. Duke. Duke. Duke.*

And just then the animal welcomed its whip-master, Mike Krzyzewski, who shouldered by me as he trailed his team onto the floor. Dressed in an undertaker's dark suit, he looked tight, cadaverous, intent. He stared straight ahead. He walked stiffly, with a bit of a hitch, like a man still aching from a bad back. The tumult now was ungodly.

I felt like a teetotaler at an open bar, around me the bright sounds of hilarity rising by the drink, and me, sober on the sidelines. Fortunately, at my moment of weakness, I spotted the funniest sign of the evening, and it was being held not by a Crazy but by a trio of fellow Carolina fans. It featured arrows pointing away from the sign toward the surrounding Duke fans, and it proclaimed simply: *"Posers."*

A few feet away, Doris Burke, the ESPN floor reporter, rehearsed for her upcoming spot. She kept looking at cue cards, mouthing the words, trying to memorize her spontaneous appearance. Nothing distracted her. The Duke band danced beside me. The cheerleaders cheered. And Speedo Guy bounced up and down in excitement. So did his pendulous breasts.

Then the game began.

Basketball is a game of angles. Speed and quickness are invaluable assets, but they can sometimes be trumped if a guy knows how to work the angles. A smart player can make up for physical deficits by mastering geometry—by knowing the quickest way between points and beating opponents to spots, on both offense and defense. JJ Redick, for instance, knew how to chop the court into productive zones by slicing crisply off picks and finding a second or two of open space, which was all he needed to launch a plausible jumper. On defense, although a step slow, he tried to cut off his man's driving angles, and better yet, to make it hard for him to even receive the ball.

By comparison, the UNC point guard Raymond Felton made geometry seem like a class you took if you hadn't already mastered the high-speed interactions of quantum physics. He was proton-fast, and with only Illinois's Dee Brown as competition, probably the fastest point guard in America. There wasn't a single player on the Duke squad who could match his speed dribbling from one end of the court to the other.

A dead ringer for the rapper Tupac Shakur, Felton was a fast talker, too, oddly for a country boy from Latta, South Carolina, where the drawl still ruled. In the player's lounge after games, his words rolled off a speedy conveyor belt, dropping into a bin marked PLATITUDE. "We're definitely going to have to play hard," he said. Or, "most definitely, we respect our opponents but at the same time, we believe we can beat anybody." He most definitely loved that word, "definitely," and it punctuated his expressions like a beat box.

He was said not to enjoy long interviews, but he spoke so fast that five

minutes with Felton was like half an hour with most guys. He occasionally looked up at reporters from his favorite chair behind the pool table,
but mostly stared at the floor or into the vague middle distance. He had
been more open and loquacious when he first arrived at North Carolina
as a McDonald's All-American who had single-handedly carried his tiny
high school to victories over such national powers as DeMatha in Maryland. Fans had celebrated his commitment to former coach Matt
Doherty as if he were the savior-to-be of a floundering program, the second coming of Tar Heel great Phil Ford.

And indeed, his freshman season, though inconsistent, had included
enough moments of spine-tingling brilliance that it appeared he might
indeed be North Carolina's salvation. When Matt Doherty was pushed
out the door in April 2003, Felton was one of the few players who
weren't jubilant.

"He and David Noel were Matt's boys," said a source close to the team.
"But in Roy's first season, Jackie Manuel was his boy. Ray and Dave felt
like they suffered the most that first year under Roy." Noel, of course,
had been injured in preseason and then forced to guard a succession of
big men. For Felton, the problem was not so much a personality conflict;
it was his conviction that he had reined in his own scoring too much,
and without receiving the acclaim that Jackie Manuel enjoyed for sacrificing his offense. The Carolina players jokingly called Manuel Jackie
"Williams" in recognition of his popularity with their new coach.

"I feel like I sacrificed my game more than anybody," Felton said with
vehemence. He had bridled under Roy's demands that a point guard
concentrate first on involving others in the offense and only then worry
about getting his own points. Looking back to the Miami game the previous season, he proclaimed, in very un–Tar Heel–like fashion, "I'm not
taking just three shots again." Summing up the previous season, he said,
"Sometimes, yeah, I took bad shots because I hadn't been taking shots."

The refinement of his game had gotten in the way of his career goals.
"I know Ray wanted to go pro after his sophomore year," David Noel
had said. Having learned how to take the bit last season, now Felton was
learning how to open the throttle again and blend his freelancing gifts
with the responsibilities of command. The results were increasingly

impressive. Even Dean Smith, with his gimlet eye and aversion to hype, regarded Felton as an extraordinary talent. "He'll rank up there with the best," Smith said. "He came along as fast as Phil. And what a great kid. He's great like Phil."

But at the outset of this game, it wasn't going very well for Felton. He appeared to be running a one-man fast break against a team that had all five defenders back. In the first three minutes, the Tar Heels threw the ball away four times, with one turnover coming as Felton streaked down the court with the ball and booted it out of bounds. He was neutralizing his own speed. It looked as if UNC was trying to deliver the knockout punch at the start of the contest, which to my mind betrayed considerable anxiety. Like Muhammad Ali leaning back against the ropes in Zaire, allowing George Foreman to wear himself out, so it appeared that Duke was going to sag back and make the game a half-court battle. If Carolina wanted to play at top speed in that sort of game, all the better for the Blue Devils.

"I was overly nervous about the game," Rashad McCants admitted later. "Players shouldn't be nervous, but I'm human. . . . I think everybody on this team was nervous, especially in the first five minutes." They had reason to be. UNC had lost three straight to the Blue Devils, the last two in particularly excruciating fashion, as if the basketball gods stayed busy inventing new methods to torture the Tar Heels and their faithful.

What was their motivation for torturing us, you ask? Well, you know how gods are. Like the U.S. Army, they have a don't-ask-don't-tell policy. But perhaps they simply wanted North Carolina fans to know that victory was not inevitable, that suffering was more common than not, that we had been spoiled by years of Dean Smith, that maybe we had gotten a little uppity and self-satisfied as partisans. And that victory always tasted sweeter preceded by parched years in the desert.

If the game so far was any indication, it looked like the gods were cooking up something special for tonight.

As the first half wore on, Duke was playing magnificent defense, continuing to prevent Carolina from getting out on its death-dealing fast breaks, trapping the Heels in the amber of a half-court game. McCants was missing open threes that he usually knocked down, and the more he missed, the more vexed he became. When Carolina managed to get the ball inside, Duke's center, Shelden Williams, was putting on a raucous block party, filling brief breaches in the Blue Devil defensive wall with his impeccable timing and reach. Were it not for Sean May's offensive rebounding and put-backs—he snared 12 rebounds and eight points in the first half—the Tar Heels would have been in real trouble. As it was, intermission arrived and they were down, 29 to 36.

The journalist in me had tried very hard to maintain control in the first half, observing himself and his surroundings with the neutrality of a naturalist hidden in a blind, studying gorillas in the Congo. But he'd suffered a couple of relapses, animal that he was.

At one point, the beast shouted "NO!" when Felton slipped to the floor and lost the ball. At another, he cried "Foul!" when McCants was mugged going to the basket.

"Cut it out," the journalist had whispered. "You can't behave this way. You only got in here because you said you were a reporter."

"I'd like to see you try and stop me," the beast said.

"At least be polite," the journalist said, who knew from experience it was best not to get into a pissing contest with the beast. "Keep your feelings to yourself. Observe them, don't act on them."

"Traitor!" the beast exclaimed. That was it for their dialogue.

I did notice one thing with cool circumspection. So complete was my identification with the action on the floor that my body tended to echo the movements of the players, swaying from one side to the other as Felton wove up the floor, leaping when Sean May crammed one in the face of Shelden Williams, holding my breath when JJ Redick fired a long three with that beautiful form. I could feel that form inside: the jump, the flicked wrist, the backspin imparted.

Unfortunately, the ramshackle first half meant that more often than not, I lurched around like a drunk trying to make his way down the hall of an ocean liner in high seas. The teams had batted the ball around the court, scrapped for it in rugby-style scrums, and clanged it off the iron to the discordant tune of a 33 percent field-goal percentage for North Carolina, 40 percent for Duke. UNC had 12 turnovers, Duke eight. The contest lacked rhythm. And that was ominous for the Tar Heels. They were accustomed to overpowering opponents with crescendos of points, having won their ACC games so far by an average of 21.7 points a contest. They craved the Wild West freedom of the open court, not the grind-it-out constrictions of half-court basketball.

The second half opened with Shelden Williams blocking two consecutive Tar Heel shots. But North Carolina cut the Duke advantage, then Duke surged, then the Tar Heels whittled the lead down again. And so it went.

Daniel Ewing sank two threes on successive possessions at a key moment in the second half, and for a moment, it looked like Duke might pull away. If a single Blue Devil had to take UNC down, Daniel Ewing might as well be the one. In the Duke–North Carolina rivalry, there were always players who stood as exceptions to the rule of universal hatred, players whom fans of the opposing school for some reason liked and wouldn't actually have minded having on their own team. Many Duke fans approved of Sean May, for instance, and given the games he was having against Duke—in fact, given his season as a whole—it was understandable that Duke partisans could imagine enlisting him as a Blue Devil. For North Carolina fans, it was Ewing who received the hatred waiver.

On the *Inside Carolina* message board one morning, posters were praising Ewing. "How good could Daniel Ewing be if he were the first option at some other school?" wrote a poster named Younsee. "The kid has got skills. If I can ask some of you to put aside the Dook bias for a second and comment, please do."

The comments came in, fast and furious. Tarmax responded: "Ewing is probably the best all-around guard in the Dook lineup. He does everything well, except, perhaps, run the point. Unfortunately for him, Dook is always fascinated by one-dimensional shooters like 'Booger' Redick, so Ewing gets relatively short shrift. He could be an All-American on just about any other team, including ours."

"The difference between Ewing and Redick is simple," wrote Goodnterribles. "Ewing is an athlete who can shoot, Redick is just a shooter."

Younsee again: "Although I may draw the ire of my Tar Heel brethren, I think he is a very similar player to Joe Forte, who was also not known for his ball handling and D, but was a prolific scorer. Just my opinion that he is under-utilized."

Tarmax chimed in: "I can see the Forte-Ewing comparison. Unlike just about any other Dookie in the lineup, I'd take Ewing as a Tar Heel any day."

4x4 Heel put it all into perspective when he wrote: "I've always kinda liked Ewing. That is to say, I hated him less than all the other dookies."

I mentioned all of this to Daniel Ewing one afternoon over at Cameron.

He laughed shyly.

I could see right away another reason why he was relatively undetested by the Carolina faithful. He possessed one of those amiable faces that in repose seemed incapable of rage. And even on the court, I had noticed that he rarely if ever played with a snarl. He actually had a habit of coming up big against North Carolina. This year, playing largely as a point guard, he was averaging close to 16 points a game, and just over three assists.

"The media and the fans might portray it like all of us guys hate each other," Ewing said. "And in fact, maybe when we step on the court, we do. But when the game is over, we say, 'What's up, man? How you doing?' You don't see a guy on the street and walk the other way."

Last season, he'd been charged with guarding the Tar Heel point guard, Raymond Felton. The two liked each other, having roomed together one summer at the ABCD camp in New Jersey, where they worked as counselors and enjoyed long conversations about basketball. Back at their respective schools, they still two-wayed each other often, and talked every now and then. "He's a cool guy," Ewing said of Felton. "I run into him occasionally at the barbershop." The barber who cut Ewing's hair was actually a North Carolina fan. He had teased Ewing "about how Raymond might just explode on you and how North Carolina has got the best five starters this year."

"You ever worry about your hair?" I asked him.

"He still does a pretty good job," Ewing said with a smile.

Ewing believed he could stop Felton. He'd done so in the past. "But it's tough. He's got great quickness. A really good spin move. If you try to cut him off one way, he'll just go the other. I just try to stay in front of him." The Blue Devils tried to keep Felton out of the paint.

Still, if Felton nailed a jumper over him, Ewing said that he might say, "Good shot." They had a mutual respect.

Ewing had arrived at Duke from Houston, where his father worked rotating shifts days and nights at a chemical plant. He was already accustomed to sharing the backcourt with a player whose star shone brighter than his own. At Willowridge High School, he'd been the sidekick to the speedmonger T.J. Ford, who'd been picked in the first round of the NBA draft in 2003 after two seasons at Texas. "That's my man," Ewing said. "We talk a lot. We just talked yesterday, as a matter of fact."

His experience with Ford had allowed Ewing to feel comfortable about coming to Duke, where he knew he might play with guys "that were as good, maybe even better" than he was. "That's why I fit in really well with the program here."

For the Blue Devils, the unexpected hero of the second half was actually turning out to be DeMarcus Nelson, who scored 12 of his 16 points in that period, including two threes, an unexpected bonus from a guy shooting 24 percent from beyond the arc for the season. With about ten minutes left and Duke clinging to a one-point lead, Nelson's angelic eyes lit up when he spotted an open lane, faked McCants, and barreled towards the hoop from the left side, banking in an old-fashioned layup.

Felton wouldn't give up, however. He kept driving the ball towards the basket, beating his defender and getting either layups or fouls. And Sean May continued his powerful inside play, ripping rebounds and nailing shots at close range.

Of course, the game was bound to come down to the last possession, just as it had in the last two contests. It was extraordinary the way this kept happening.

After Rashad McCants finished a rare North Carolina fast break, instigated when Marvin Williams blocked a shot, Duke took possession with just under a minute left in the game, clinging to a one-point lead at 71 to 70. With only four seconds on the shot clock and the Crazies shrieking, Redick hoisted a 30-foot jumper from straight away. For seven straight games, he'd been on a tear, scoring 20 or more points each and twice going over 30. Tonight, shadowed by Jackie Manuel and David Noel, he'd made 18 points so far, hitting four of ten from three. And 18 is what he would still have at game's end. The ball soared into the air, appeared to tire in midflight, and fell to earth short of its target like a dead duck. Yes, right there in Cameron Indoor Stadium, where back in 1979, during another Duke–Carolina game, the very chant of "air ball" had supposedly been invented (imagine a world without it), JJ Redick, the long-range assassin, whiffed. That is to say, he shot an *air ballllll, air ballllll. . . .*

The ball came down over the end line, and North Carolina immediately called time-out.

Silence.

Speedo Guy crossed his arms, shivering like a kid by the side of a pool.

Eighteen seconds in a Duke–Carolina game was long enough to constitute a geologic epoch. In the Tar Heel huddle, Roy Williams was calling for his team to run the Long Beach set, a play designed with five options that the Heels had successfully employed to beat number-one-ranked Connecticut the season before on a three-pointer by McCants.

This would have come as no surprise to the Duke bench. Krzyzewski knew the play was one of North Carolina's favorites at such moments. The day before, the Blue Devils had walked through the Long Beach as they rehearsed for end-of-the-game situations. They had also watched the set on videotape. "Even if we knew they were going to run it," Krzyzewski said, "they're good enough to execute it and still beat you. Someone still has to make a defensive play."

In the huddle, Krzyzewski told his team, "Watch for McCants."

The two teams were now on a course for direct collision.

From my seat under the UNC basket, I watched the play unfold. Sean May inbounded the ball to Felton under the far basket. He sped down the court. Daniel Ewing picked him up. But as he had throughout the second half, Felton spun and got a huge step on Ewing. The space between Felton and the basket yawned like a stretch of empty highway. All he had to do was drive it. He would arrive at the basket within a second or so. Even if Shelden Williams, who was playing goaltender, arrived at the same time, chances were excellent that Felton would get a close shot, a foul, or an open passing angle to a teammate suddenly free because of Duke help-side defense.

But for some unfathomable reason, Felton froze. He picked up his dribble. The beaten Ewing was now back in the play, harassing the UNC point guard, making it hard for him to do much of anything. The stretch of open highway had turned into a roadblock, with Ewing policing his every move.

The key to the play: Felton had been focused on getting the ball to McCants, a plan that was foiled when JJ Redick, hardly well known for his defense, had blocked McCants as he attempted to come around a screen. The shooter who'd just shot an air ball had redeemed himself with a clutch defensive move, thwarting McCants just as he had the previous season at the end of the second regular-season game between the two teams.

With the clock ticking inexorably towards zero, Felton weakly pushed the ball to his right towards a flummoxed David Noel. The Durham native chased the pass, trying to corral it, but lost it out of bounds as time expired.

The noise came in spiraling waves, mounting one crescendo, then another. It was first the sound of relief, then of exultation.

I stood silently at the edge of the court. Redick and Lee Melchionni embraced and raised their fists towards the crowd. ESPN's Doris Burke interviewed Redick, who'd played all 40 minutes. Krzyzewski stepped lightly by, an unmistakable smile playing across his face. Finished with the interview, Redick lifted his arms to the students as he bounded to the locker room. The North Carolina players could hardly move, wandering the court as if trapped in a bad dream of their own devising. Felton and McCants stared at each other long and hard, never breaking expression.

In the media room, the press was buzzing about the final play. Andy Katz of ESPN was saying: "I can't believe they didn't get off the shot. Felton had been going to the basket the whole game. He made that turn . . . and . . ." And nothing. Duke hadn't scored a single field goal in the game's last five minutes and had somehow still won. Carolina had outrebounded the Blue Devils 43 to 28 and had somehow still lost.

Of course, 23 turnovers, eight of which were Felton's, can do that to a team. So could Jawad Williams's shooting one for six from the field, McCants's going three for 13.

"We would have gotten beat by 40 without Sean May," Roy Williams told the media. "I think the whole thing boils down to, Duke made plays and we didn't."

The Duke locker room was jubilant.

The visitor's locker room, by contrast, was the kind of antique, dingy place where Knute Rockne might have given a speech to a bunch of white guys in leather helmets. The Tar Heels huddled there like refugees from a cataclysm. Quentin Thomas, who'd subbed briefly for Felton, sat in a plastic chair, staring into space, rubbing his feet. The assistant coach Steve Robinson slumped next to the Coke machine, lost in thought. Reporters hovered over Felton, still in his uniform half an hour after the foiled last play. "I should've took it," he said. "It was a mistake I made. I was looking for Rashad on the other side of the screen, so I took it back out. He wasn't open, and Sean wasn't open on the slip. So the whole play just got terminated, just like that."

Sean May dressed by his locker, voluble as ever, even in defeat. "It's a great rivalry," he said. "Shelden Williams is a friend of mine, and it was a terrific battle with him. There's a ton of things we can correct. We're gonna break down the game film. Our season's not over."

AFTER THE GAME

The rain started right after Duke won, as if God were directing a cheesy TV movie of the whole affair for the Lifetime Network. My ears were still ringing as I drove home. The din inside Cameron after Carolina failed to get that shot off had been terrible.

The most poignant sight of the evening came as I left the dingy little visitor's locker room and climbed the dark stairs. Alone on a step sat Roy Williams, head in his hand, slumped in abject desolation. He looked like the statue of *The Thinker*, if what *The Thinker* was thinking about was why Raymond Felton had broken off his drive to the basket after getting a step on Daniel Ewing and finding himself wide open at the top of the key. He didn't penetrate and he didn't shoot. What was Ray not thinking?

The contest had revolved around momentous little things like that. It was an ugly game in which either will or luck had prevailed—or some tangled combination of both. It revealed how fine the line was at the highest level between success and failure.

Maybe that was why Roy Williams was alone in the stairwell. He might have been wondering whether chance had ruled against him, whether he was cursed against the Blue Devils, having as North Carolina head coach lost three consecutive games to them on the last possession. I knew Williams owned a superstitious streak. So he must have been thinking about the perversities of his mojo. Or maybe he suspected that in some mysterious way, the Tar Heels hadn't wanted the victory quite as much as the Blue Devils. Or maybe they had wanted the win *too* much. Maybe Roy Williams was thinking about why Rashad McCants couldn't hit an open shot. Perhaps he was blaming himself for calling the Long Beach set. It had worked so well against Connecticut a season ago, but Duke seemed to have been prepared for it. All these were unsettling possibilities that could insinuate themselves into a coach's mind like invisible toxins drifting at night from the local chemical plant.

Still, as bad as Roy felt—and he looked cadaverously bad—at least he enjoyed the clarity of knowing exactly why he felt so bad and exactly what he could do to feel better the next time. Which was win. And that was one clear good thing about sports. Life was usually murky, glimpses of light through a dirty car window on a fast drive. But games inscribed a clean and simple line between the light and the dark. You won or you lost. Good, bad. Light, darkness. Sleep, insomnia.

I did the math: Carolina had 23 turnovers, which was way too many in any game, especially one in which Duke ran long offensive possessions. Rashad McCants nailed only three of 13 from the floor. Sean May, at least, enjoyed a monster game against the Devils, 23 points and 18 rebounds. To lose a game by only one, at Cameron, when North Carolina was playing below its capacities—that stood as a perversely good omen. Maybe that's what Roy Williams was thinking. But I doubt it. Not on the stairwell. Maybe tomorrow.

I came into the house and my mother was in her nightgown. The first thing she said to me was, "What was he thinking?"

I knew exactly what she meant.

"I don't know, Mama. He had the corner turned. Ewing had gone for the steal and missed. I just don't know."

"That is a sad way to lose," she said.

"I know. The worst. Not even to get a shot up."

"What was he thinking?" she asked again, almost pleading. "That was terrible. He was open."

"I know," I said. "I know."

"He'd been going to the basket hard in the second half," she said.

"I know. I agree with you. He had him beat. I don't know why he didn't drive. He would have gotten fouled, or got the shot, or found an open guy when they doubled him."

Some questions, it is clear, have no answer. Does God exist? Will there be enough Social Security for us when we hit retirement age? And what was Ray Felton thinking when he picked up his dribble at the top of the lane with only seconds left in the game and nowhere to go with the ball?

"I've got to get clothes out of the dryer," my mother said. It was after midnight and she didn't usually do this after midnight. But I understood; fulfilling a routine task can help you get over this sort of loss.

"Do you need any help?" I asked.

"I've got it," she said.

The rain came down. And those frogs in the swamp kept peeping. Fools.

EIGHT

Danger

YOU COULD BE IN BIG TROUBLE

WITH A GLASS EYE that was staring at me with the implacability of the mineral world, Professor Robert Thurman looked like a cross between a mythical Tibetan monster and a doctor of religious studies, the latter of which he actually was. On an emergency mid-season trip to New York, I stopped by to see him at Columbia University. I had to know from the point of view of a renowned scholar and practitioner of Tibetan Buddhism whether hatred for Duke might cause me to be unduly reincarnated, forced to spend billions of years as a praying mantis or a screech owl or a coyote baying at a coldhearted moon.

Baying seemed an especially apt fate.

"Dr. Thurman, I know you believe in reincarnation," I had said after our rapid formalities were dispensed—he was harried, the phone was ringing off the hook, the Dalai Lama was soon in town, students kept popping by, asking to be put on the guest list for "His Holiness" (as one brownnoser put it), and he had to teach a class shortly. "I need to know," I told Thurman, "what could happen to me if I keep going like I am. My problem is that I am occasionally overcome by hatred for a particular basketball team."

He fixed me with the eye. One of them, anyway. For such a doctrinaire Buddhist—he could read you the letter of the karmic law—there was nonetheless something swashbuckling about Thurman's self-presentation. He was wearing a denim shirt, the sleeves rolled up to reveal a stevedore's forearms. He combined baleful scrutiny with a wicked grin. And like a monk of the Dalai Lama's Gelugpa sect, he seemed to enjoy disputation. Yet all around his small office on the second floor of the religious studies department overlooking Claremont Avenue was the evidence of an immersion in deep contemplation, the study of the mind, the pursuit of peace. Tibetan texts wrapped in orange and purple cloth sat on several shelves. Another bookcase was lined with volumes of psychology, Daniel Goleman's *Destructive Emotions*, for instance; books of philosophy; Robert Nozick's *Anarchy, State, and Utopia*; and hundreds of classic tomes about all varieties of Buddhism.

"What happens to you depends on what kind of hatred you're talking about," Thurman answered. "There are many different kinds. If hate means entertaining a negative judgment about something, like, I hate this kind of ice cream, that's perfectly okay."

"It's not ice cream," I said.

Still, I felt a glimmer of hope. Like a debater, I had my arguments lined up. Is it wrong to hate something hateful? What if you hated injustice? What if you hated *symbols* of injustice? "So hatred doesn't necessarily have negative karmic consequences? In fact, it can sometimes be a good thing?" I said.

"Right," Thurman answered. "In fact, you have to be negative and critical about some things, because some things are no good. That's part of ethical judgment. Wisdom is defined as the knowledge of the difference between good and bad. It involves discrimination."

The key thing about hatred, he said, was when it evolved into a *klesha*, an addictive emotion, "something that grabs ahold of you and twists and tortures you. The emotion is in command of you. You have to follow its dictates. It purports to help you either in getting something you want or destroying something you don't like. When hatred involves harming others, based on your having become a tool of an unreasonable energy, that is when it becomes a *klesha* and generates negative actions."

"Let me give you a scenario," I said to Thurman. "Let's say I was sitting in a room, watching a basketball game on TV, and in a fury, I began to scream at one of the coaches on the screen. Would that constitute bad hatred?" I wasn't speaking only for myself here. There were millions of fans who behaved this way. Maybe I would get some points for clarifying things for us.

"That sounds bad," Thurman said. "Sounds exhausting. Sounds useless, because they can't hear you." He grinned devilishly at the image. And didn't I know it—all these years of yammering at the TV, the way my father used to talk back to Jesse Helms during his "Viewpoint Editorials" for WRAL-TV. We were fools, plain and simple. And evidently destructive as well. "And if anyone else were present," Thurman added, "it would inflict an unpleasant vibration on them."

In that case, I had definitely inflicted a lot of unpleasant vibration in my time—on a wife, girlfriends, children, dogs, and cats. They often left the room in the wake of my "vibration." My Carolina friends were different. Together, our unpleasant vibrations had the sweet ring of friendship, united in our love for one team and hatred of another. "Yes," Thurman concluded, "that kind of behavior would be considered something very negative."

It seemed I was suffering from a *klesha*. Thurman appeared a little worried for me, like a doctor who had spotted a shadow on the X-ray. "The Buddhist project is to conquer the unconscious, not just the conscious," he said. "The only way to do that is to realize the limitless consequences of what you do." He added a proviso that seemed tailor-made for me: "You can be fierce with people if your motivation is friendly and loving. But hatred—it's hard to imagine good hatred. Because it just wants to destroy what it sees."

"There's a primitive-magic idea that sports fans believe in," Thurman said. "Like if you do certain things, you're giving your team power and energy."

I thought of my friend and fellow Tar Heel Mike C. and his magic key chain, which he brought out only during crucial moments in the game, shaking it at the television during an opponent's free throws. The key chain boasted a success rate of at least 50 percent. Was Thurman telling me that the key chain didn't really work?

"You seem to feel that being a sports fan of any sort has seriously negative consequences," I said. "But especially if you are a Duke hater."

He gave me a look that said I better not be having too much fun with all of this hatred. "In a community of fellow fans," he said, "you're mutually bonding over your animosity and your allegiance. But maybe you're giving the team and other fans the wrong kind of power. In a militaristic society like ours, team sports like football and basketball, where you're supposed to vilify and demonize the enemy as a player and a fan, train you to demonize an enemy. And it's like when they say, you have to come and attack Iraq, you're likely to think, 'Oh, great! Let's go attack Iraq!'"

Militaristic! *Demonize*! Such frightening words! "I hate to think that every time I yell for a ref's head, I'm dispatching our boys to Iraq," I told Dr. Thurman.

I was reminded of my father shouting with exasperation, "Throw in the Christians!" He recoiled against that sort of mass bloodlust. It had seemed a little prissy of him at the time, like a guy who went to a dance club and asked everybody to stop doing the monkey.

"Let's get to the bottom line," I said. Time was running out. Thurman had to scrutinize an old Tibetan text in preparation for a class he was about to teach. I asked him the sixty-four-thousand-more-eons-in-karmic-returns question: "So, karmically speaking, it could be bad news for me if I kept hating Duke?"

"You could be in big trouble," he said. "It is endangering. And you really shouldn't be overidentifying with a winner and loser to that extent." He was smiling compassionately, even chuckling.

I must have looked gloomy. To cheer me up, he said, "Fear of endless lives of suffering is motivating. You've got to do something one by one about all the bad things you've done that are following you."

"That could take a long time," I said. "I'm not sure I have that much time."

"Never give up," Thurman said with a pranksterish smile. "Because it can always get worse."

When I visited my girlfriend later, I told her about Thurman's warnings. I wondered what miserable life form I might return as.

She wasn't worried about herself. "I'm coming back as a dolphin," she said. "Or as the beloved lapdog of a doting and wealthy homo."

"I should be so lucky," I said.

A ONCE-VICIOUS RIVALRY
UNC versus NC State
Raleigh
February 22, 2005

Rashad McCants was out for tonight's game at State with a mysterious intestinal ailment, from which it would take him most of the next month to recover. In his place, Melvin Scott would start. This was the sort of break that Melvin had been hoping for, the chance to show what he could do if given the minutes. The game began, and it wasn't long before Melvin arched in a three from out near Fuquay-Varina. The shot appeared to buoy Melvin's confidence.

The first minutes of the game were hard-fought, as befitted a once-vicious rivalry. In the Seventies, while the Duke basketball program slept, North Carolina State served as the archfoe for the Tar Heels. In 1973 and 1974, State fielded two of the best teams in ACC history, both of which starred probably the greatest player ever in the league, the six-three small forward David Thompson. The Wolfpack went 27 and 0 in 1973, but missed the postseason because it was on NCAA probation for the improper recruitment of Thompson. (A Wolfpack booster had paved the Thompson family's dirt driveway, an apparently irresistible enticement in the eyes of the NCAA.) The following year, State went 30 and 1 and finished the season as NCAA champions, derailing UCLA's unprecedented seven-year title run.

The psychological dynamic of that rivalry was oddly analogous to the current war between Carolina and Duke, only the roles were reversed. NC State played the role of Carolina, the humble state university with an illustrious b-ball history. Carolina took the part of Duke: snotty elitists looking down at their tractor-driving cousins who'd been forced to attend that second-class institution in Raleigh.

Now, decades past their glory years and filled with class resentment, the Wolfpack fans howled with fury every time Carolina came to town. UNC was the lover that had moved on and up, State the dumpee that had downsized into a dumpy apartment, where he spent the weekends microwaving frozen dinners and watching network TV. You witnessed the bitterness last season in Raleigh when Rashad McCants went to the free-throw line with the game hanging in the balance, there to be greeted by an unusual form of invective. "STD, STD, STD," chanted the State fans, accusing McCants of having a sexually transmitted disease. McCants's face cracked into a disbelieving grin. It seemed that the State fans had hit upon a magic incantation. They couldn't have inspired McCants any better if they'd screamed, "We love you, Rashad." He calmly sighted his free throws and bottomed them, contributing mightily to an important away victory for a team that was road-challenged that season.

In tonight's game, State clung to a narrow lead for much of the first half. With close to six minutes left, a three by Reyshawn Terry cut the Tar Heel deficit to one. Then Melvin canned another three to give Carolina its first lead in ages at 28 to 26. He was performing marvelously, picking his spots on offense and playing tough defense. At one point, he grabbed six-ten Jordan Collins and held him. This may have been a desperation move, and Melvin got whistled for the foul, but it was still inspiring to watch a six-one guard muscle a big man. Near the end of the half, State's Julius Hodge sank a three for the first time since the Carter Administration; it had to be against us, I thought. It appeared Carolina would go into the locker room up by only two, 35 to 33.

But at the buzzer, from deep in the corner, Melvin struck with a last-second three to nudge the lead to five. He loved those shots. He had no time to think. I vaulted into the air with joy.

At the start of the second half, Melvin picked up where he left off, nailing yet another three. The shot fell down from the upper atmosphere like a piece of space junk. Look out below! He ran down the court with a swagger, and it looked like he was gnawing a bone between his teeth, growling. Hepped up, he dived to grab the ball as State bobbled it on the baseline. By this point, he had done his job, filling in admirably for McCants.

The game died a slow death. Hodge, a weak outside shooter, hit several more threes, and State kept Carolina from ever stretching its lead much beyond ten points, which in fact represented the final margin, 81 to 71. The once-rabid State fans filed out in silence. Their team was on the good side of mediocre, but no better. And there were no signs of that changing anytime soon.

The next day, I called Melvin. The outgoing message on his cell phone seemed apt. "Never give up," Melvin intoned. "You only beat yourself when you stop trying."

LET IT RAIN
Practice
The Dean Dome
February 24, 2005
3:40 P.M.

When I arrived for the North Carolina practice today, I ran into A.J. Carr, the elfin, silver-haired reporter who'd been at the Raleigh *News & Observer* since 1966. He was the courtliest of men, well liked by coaches, players, and colleagues alike. He in turn admired Dean Smith, who on a road trip to Clemson had scribbled plays on napkins for Carr to use in his season as a youth-league coach. Carr liked Krzyzewski, too. "He's intense, but he's nice when he's not on the court," he had told me.

"What are you doing here?" I asked.

"I'm on the Rashad McCants stakeout," he said. "And he is nowhere to be found." That was the mystery of the moment: Where was Rashad McCants and what was wrong with him? A doctor who worked at UNC told me that more than three hundred people had tried to access McCants's private medical records.

Turned out that this afternoon's practice was just a shoot-around. The players had lifted weights earlier that day. Assistant coach Joe Holladay ambled from player to player, watching the shooter's form. Melvin came out and started his routine, jumpers near the basket, then farther out. But no Rashad.

Holladay stopped by David Noel to demonstrate how the junior swingman should bend his knees at the free-throw line. They were both being watched by Bill Guthridge. The diligent sounds of sneakers squeaking and basketballs whapping on the floor was broken when Melvin stripped off his shirt, took a long three, and yelled, "Let it rain!"

Then chaos broke out on the floor. The reserve C.J. Hooker emerged from beneath the stands with a water rifle strapped to his arm. He pumped it a few times, then blasted a long stream of pressurized water directly into the chest of Joe Holladay. Melvin kept shouting, "Let it rain, let it rain!" Now I saw that he wasn't referring to long-range jumpers.

Holladay took his dousing well. He feinted at Hooker, who retreated to the sidelines and tried to conceal his weapon. Holladay assigned managers to mop the floor with towels. It was one thing to be blasted in the chest by a water pistol; that, a coach could survive. But it would have been a setback of an altogether different order had a player on one of the top-ranked teams in America slipped on a wet spot caused by a squirt gun and wrecked his knee.

As if nothing unusual had happened, the team went back to shooting. Holladay engaged Melvin in a three-point contest. "You should be embarrassed," the coach told him at one point, scoring a point by sinking a deep-knee-bend set shot that looked like it hadn't been used since the 1940s. But the technique worked. Bending those old but boyish-looking knees, flicking the ball netward, Holladay won the first round in the best-of-five contest. Melvin barely made a victory in the next two games. In the fourth, he caught fire and torched Holladay. Melvin whooped, and Holladay moved on to another shooter.

Quentin Thomas came up behind Melvin and just stood there, breathing on his neck, goofing on the man he called his uncle.

BRING ME THE HEAD OF ROBERT E. LEE

While searching for a book the next day, I discovered under my bed a bust of Robert E. Lee. "Where did that bust of Robert E. Lee come from?" I asked my mother.

"Robert E. Lee?" my mother said. "Where?"

"Under the bed," I said.

"What?" my mother said with alarm. She was perturbed at the thought of Robert E. Lee anywhere in the house, especially under the bed.

"No, not Robert E. Lee himself," I said, "a *bust* of Robert E. Lee."

"Where?" she asked.

"Under the bed," I said.

"I have no idea," she said.

However he got there, wherever he came from, he was down there. Under my bed. In a little white box filled with tissue paper. A little metal head of Robert E. Lee. You can run, boys, but you cannot hide. The metaphorical possibilities for this were endless.

The head–of–Robert E. Lee episode reminded me of how my mother had become deeply Southern, despite having started out somewhat behind in that respect (Scottish-Lebanese family; Northeastern upbringing). She once told me that she liked everything about the Civil War documentary by Ken Burns except how he treated the Southern generals. I listened in astonishment. "Why do you say that?" I asked.

"He was too hard on them," she said.

I had no idea.

Where my father found home in the place he started, my mother discovered it in the place she ended up. Her family were voyagers, having gone great distances in the last generation, that old American song. Her mother was a Scottish immigrant who crossed the North Atlantic with her family as a nine-year-old the same week the *Titanic* went down. Her father was of Lebanese descent and got his start selling bootlegging equipment during Prohibition, the first display of a knack for mechanics and business that eventually blossomed into a career as an engineer, constructing buildings and highways all over the continent.

And then it was my mother's turn to hit the road. After she finished at Mount Holyoke, she went to work for a doctor in New Haven, and when he decided to accept a post at the University of North Carolina, he asked her to come supervise his laboratory team.

So right around the same period that Everett Case showed up in Raleigh to ignite the state's obsession with basketball, my mother arrived in Chapel Hill. The South was more Southern then. She might just as well have traveled an Old Spice Route and ended up at the palace of Kublai Khan.

The marvels she beheld: Humidity shimmering like the mirage of a mysterious green city. Bugs the size of shuttlecocks swatting against the screens. The roar of june bugs in the vaulted heat of summer afternoons.

Negroes. The radio sweltering with gospel and Pentecostal preaching. The medical-school dean with the Confederate flag hanging on his office wall. And the easy, roundabout manners of the doctors at the hospital, who teased and drawled and got to the point they introduced on Monday by about Wednesday or Thursday, who seemed to own time as if it were an old family estate. One of them would become my father.

She liked grits the first time she ate them, at the Carolina Inn during her first weeks in town. Her family up North didn't even know grits existed. When her brother came down for a visit, he exclaimed to my mother, "God, how can you eat these? They've got no taste." She liked them, she told him.

She was alone in the new world, but she had a job and, in Dr. Welt, a boss who looked out for her. Her father had sat Dr. Welt down before her move to make sure that she was in good hands. And like most women, she didn't look backwards with longing—unlike my father, who in time became nearly bent backward by his recollection of a life that existed most perfectly on the private grounds of memory.

He saw her in the hospital cafeteria; he had been watching her a while, the story went. And one day he came over in his white doctor's robe to introduce himself. He was polite, charming, proper enough, but with a twinkling subversion in his eye. I knew that look. They went to a movie at the Varsity on a double date, and after the movie my father suggested that both couples drive to the beach right then and there. My mother was game, but the other man grew flustered and suggested that they go home and sleep and drive to the beach the next day. The other man was something of a mama's boy, my mother said years later.

So through marriage, four children, working at the hospital as a laboratory technician, and weekends with my father's family in the paradise of Huntersville in Mecklenburg County, she became by easy degrees a North Carolinian. In my father's last years, she even began attending meetings of the North Caroliniana Society with him, and she continued going after his death. This was her place now, its history her own. Had she been like my father, this never could have happened. He compared the rest of the world to North Carolina and found it lacking. She found a reason to stay in this place and made it home.

NINE

A Spy in the House of Hate

TOWARD THE END OF FEBRUARY, I visited an undergraduate journalism class at UNC, where I'd been invited to speak about my igno-minious career in magazines. In the midst of a freewheeling discussion of journalistic ethics, I wondered aloud whether objectivity might not be an overrated virtue. The instructor blanched. What I meant to say but didn't say very well was that the worst kind of bias often hid inside the boxy gray suit of objectivity. And that the disguise could be worse than the concealed prejudice.

The students asked me when I was going to Duke.

I had told them about my experiment in crossing over to the dark side for a sustained bout to see if I ended up in the same place where I started: as a raging, bilious Tar Heel fan. Or whether I might come back a changed man.

Next week, I told them. I'll test my fanaticism. Maybe I'll actually like the folks over at Duke (wink, wink). That might be disastrous for my worldview. But in the name of unbiased, objective journalism, I simply have to find out.

After the class, one of the students stopped me in the hall. "Don't let those Dukies snow you," he said in a North Carolina country drawl. "Don't get too objective on us."

"You're a quick study," I said. "I'll do my best." We laughed. We thought it was all very funny.

MY EX-WIFE'S DUKE FRIEND

Despite Duke's prominence in my inner life, I had somehow managed to make it this far without encountering more than a few of its students or graduates, not that this had ever stopped me from bald and passionate assertions about such types.

There was, however, my former wife's friend—let's call him Ben Languid—who'd gone to Andover and then Duke. In the early Eighties, he visited us once at UNC, where we were boyfriend and girlfriend. Lounging like a sultan on my gal's Indian bedspread, he exuded a casual disregard for rules that seemed aristocratic, proffering me a lit joint with the door to her dorm room wide open and freshmen from little country towns peeking in at our combustion.

He and my girlfriend called each other by nicknames. Everyone at Andover, it seemed, had nicknames, usually with a masculine bite, even if they were women. Hers, apparently, was J.K. Ben Languid's hair spiraled out from his head in exasperated tangles, as if recoiling from whatever was happening inside his skull. He wore glasses that should have made him look helpless but didn't. Ripped jeans. Scuffed Chuck Taylor high-tops. He didn't take sports at all seriously, and laughed when I asked him about the prospects for Duke's basketball team. His sport was the Grateful Dead.

If he eventually became a Cameron Crazy, which I doubt, it would have been for the experience of being a Crazy, for the theater of it, not because his psychic survival was at stake in an epic battle between good and evil. He was just passing through the area. Being a Crazy would have been a very good joke for a few years of school, undergraduate and professional, then off to practice law and smoke a joint in the bathroom of his sumptuous New York apartment with the ventilator fan humming so the kids wouldn't smell a thing.

But my impression of Ben Languid was 20 years old, and in any case a thin basis for incriminating an entire student body. I needed to go among the enemy where they lived, to know their secret thoughts and take the full measure of their wicked ways. So I arranged to meet Mike Hemmerich, one of the founders of the *Duke Basketball Report*, an Internet site known to friend and foe alike as the *DBR*.

HATE IS GOOD

Mike Hemmerich and I were standing on the seventeenth floor of the only skyscraper along Highway 15-501 between Chapel Hill and Durham. Known locally as The Green Weenie or the Phallic Tower, the building just happened to be Hemmerich's baby. That is, he and his partner, a former Duke quarterback named Anthony Dilweg, had recently purchased the Phallic Tower for their commercial real estate business. We stood at the top of his skyscraper, staring like a couple on a cruise ship at the undulant green horizon.

"So this is your baby now," I said.

"Yeah." Hemmerich laughed.

"Not bad."

"Our little icon," Hemmerich said.

"Always nice to have a little icon," I said.

"Possibly some ego involved in trying to buy it," he said.

He and Dilweg had originally been outbid. "How bad do you want this building?" Dilweg had asked his partner. "Real effing bad," Hemmerich had answered. They promised the seller to round up an extra million in five days, and they did.

Before we sat down to eat in the University Club, we gazed a little longer over the rippling green sea of pinewoods, as if confirming the beauty of the deal that led to such a view. Even I felt momentarily like a real estate magnate. Underneath the canopy of trees lay the fabled sunken cities of Bojangles, Midas, and Bob Evans. But from up here, the evidence of franchisedom was mostly invisible. It was possible to imagine that old Triassic sea surrounding us, just as my father had described it to me when I was a boy.

We ate roast beef and new potatoes and green beans in the dining room of the club. Thick white tablecloths, heavy silver. The occasional burst of laughter from a nearby table. Now well into his forties, Hemmerich was a congenial guy who'd grown up in a small town outside of Reading, Pennsylvania. He'd entered Duke in January 1977 and fallen in love with both the school—he had ended up with three Duke degrees—and the basketball team. He'd taken a course called "Coaching Basketball in Secondary Schools," taught by Bill Foster, then the head coach, and witnessed strange marvels, like the Duke–Carolina game at Cameron in February 1979 that featured the legendary halftime score of 7 to 0. (This partial score was perhaps more widely known than the final, 47 to 40, in the Blue Devils' favor.) Dean Smith had elected to slow the game down because several days earlier he'd watched Clemson hold the ball against Duke and prevail, 70 to 49. And Smith was always confident in his delay game, which usually spread the court to UNC's advantage.

"UNC took two shots," Hemmerich said. "Both by Chickee Yonakor."

"Don't remind me," I said. In fact, one of the shots may have been a desperation heave at the half by the Carolina guard Dave Colescott. But Yonakor certainly deserved the lion's share of notoriety, since his miss was first.

"And both were air balls," Hemmerich continued. "And that was supposedly the birth of the air ball chant. During that game, when UNC was holding it, we started chanting, 'Bo-ring, bo-ring.' Then when Yonakor shot the air balls, we went, 'Air-ball, air-ball.'"

So here was a Duke graduate claiming to be present at the invention of the air ball chant. Maybe so. I wondered if this culturally significant moment could be verified, in the same way that scholars had recently tracked down the guy who yelled, "Judas," at a Bob Dylan concert in 1966.

These days, Hemmerich was channeling his passion for Blue Devil basketball into the *DBR*, where his cyber-tag was Boswell. The site had its origins in a bulletin board on Prodigy back in the mid-Nineties where Hemmerich and another Duke fan, Julian King, used to post about basketball. Mingling online with Duke basketball fanatics had been so much fun that after Prodigy shut down, Hemmerich and King decided to create a site of their own on the Internet. The first month of its existence the *DBR* received six thousand hits. Now, said Hemmerich, it averaged over a million hits a day.

He envisioned the *DBR* as a neighborhood pub in cyberspace, and joked that sometimes when Carolina fans came calling, he had to act as a bouncer. In fact, it was impossible to enter the Duke site via links occasionally posted on *Inside Carolina*. Access for obvious partisans had been blocked at the outset. Once, Hemmerich had posted a picture he'd taken at a Duke function of his three-year-old daughter asleep at his and his wife's table. He'd playfully positioned an empty wineglass by his daughter's head. The next thing he knew, the shot had been reproduced on North Carolina and Maryland websites. "They talked about how sick we Dukies were, pumping alcohol into their kids, making them pass out," Hemmerich said. He laughed. "I'm like, 'C'mon, guys.'"

"Would you say that the atmosphere surrounding the Duke–North Carolina rivalry has changed since the days you were in school?" I asked. I was under the impression that the late Seventies had been a relaxed epoch for Duke basketball, despite the 1978 team that had made it all the way to the national championship game before losing to Kentucky and Jack "Goose" Givens. I remember that I had even considered (for about two seconds) rooting for Duke in that game. After all, weren't they a North Carolina team? Nah.

"No," Hemmerich said. "The intensity was there then. You have this purity of hatred." He laughed happily. He admitted that Krzyzewski's challenge to Dean Smith in the Eighties had heightened the animosity. But it had always been there, he said. It had never gone away. And he loved it.

Purity of hatred! It sounded as refreshing as bottled spring water. I felt as if I had discovered a comrade in arms, even if his arms were pointing directly at me.

"I always thought that the hatred runs more from UNC to Duke. But you're feeling it, too!"

Hemmerich qualified his stance slightly. "I think we hate each other equally," he said. "But I think Carolina is probably more obsessed with Duke than Duke is with Carolina."

As an example, he cited how in 1999, when Duke lost in the national finals to Connecticut, Carolina students flooded Franklin Street in celebration. When Carolina was defeated in the NCAA semifinals by Florida the next year, Duke students did, well, nothing.

He wrestled with how to phrase his next point. "I don't mean for this to sound . . . well . . ." He never finished the sentence but I saw what he had left out. "Duke has higher admission standards than Carolina," he continued. "And I think this creates an inferiority complex in Carolina people. They automatically get defensive. But Duke people don't get defensive. They can call us the University of New Jersey at Durham, but it doesn't really hurt."

Oh, but how it stung me. My worst suspicions were confirmed. I had feared Duke noblesse oblige. Snobbery. Call it what you may. I had feared an asymmetry in the hatred between Duke and North Carolina. I had feared that the Duke fans might not hate us as much as we hated them. There was nothing worse than hating people and not having them reciprocate your hatred. Perhaps such people were so successful that they could afford to ignore you (the way Tar Heel fans could now ignore North Carolina State). Or maybe your victories meant nothing in the larger scheme of things; maybe the Dukies were like those Ivy League students chanting in the stands as their university succumbed on the football field to some burly state school: "That's all right, that's okay, you're gonna work for us some day." Here was class rearing its well-groomed head as the ultimate rebuke. You may win the game, but you will lose the game of life.

"Did you know," Hemmerich asked, "that there are more North Carolina graduates in this state than there are Duke graduates in the world?"

"You like that feeling of being encircled," I said. "Of us against the world." Krzyzewski liked that, too.

"I do," he said, smiling.

I should have hated Mike Hemmerich. But on this sunlit afternoon high above the pines, it seemed I was suffering a failure of my inner beast. It seemed Hemmerich was a pretty good guy. And he even bought my lunch. What was I supposed to do, punch his lights out?

Hemmerich sipped a cup of coffee, surveying the lingering diners in the elegant club on top of the Phallic Tower. He was in an expansive mood. "You know that line in the movie *Wall Street* where Gordon Gekko says, 'Greed is good'?" Hemmerich said. "Well, you know what? *Hate* is good! Because what it does—it creates this common bonding experience."

I laughed. I couldn't have said it better myself. On that, we could agree. We were fellow members of the Brotherhood of Hate, Duke and North Carolina divisions. With nods of recognition, we shook hands and promised to keep in touch.

THE TOUGH JEW

At the start of my foray behind enemy lines, I went up to New York to visit Art Heyman, an All-American at Duke back in the early Sixties. I had spotted Heyman—I presumed it was him—several months earlier through the plate glass window of his bar, Tracy J's, on 19th Street in Manhattan. There in the window in the middle of the afternoon sat a big man wearing a broad-brimmed leather hat. He was reading a Blue Devils media guide for the upcoming season. The bar was empty. This sight must have been vouch-safed me for some reason—a parable about the transience of fame? A museum-quality diorama of the posthumous life of a former All-American?

I knew that Tracy J's was Heyman's bar because above the front door hung a sign that all but shouted the essential facts: *Art Heyman Owner Former NY Knick Duke University All American College Player of the Year.* It seemed a little poignant for Heyman to be advertising himself this way in New York; if he had to say all that so that people knew who he was, maybe it was better not to say anything at all. I could better envision such a banner festooned above a watering hole back in Durham into which alumni might wobble after games and pretend to have been buddies with Art. *Honey, this is Art. Art, this is my wife, Carol. Now, Art, big man, don't tell Carol about the things we used to do back in our wild days, heh heh heh.* However, Heyman's ego was said to be the size of a thin-skinned king's, and a big sign in the New York street proclaiming his resume was certainly more visible than a yellowed news clipping in a frame above the bar.

I walked in to introduce myself. "Are you Art Heyman?" I asked, extending my hand. He studied it for a long moment before shaking it. I seemed to have intruded into a moment of private grievance. "Who are you?" he wanted to know. When I told him I was interested in his days at Duke, he immediately handed me a Xerox of a box score of the famous 1963 game in which he scored 40 points against North Carolina. He just happened to have this lying around.

"Thank you," I told him.

"I'm not giving it to you," he said. "It was my last game against Carolina. I should have scored sixty."

"What happened?" I asked.

"I was nervous. My last game. I missed a lot of easy shots." He pointed across the bar to a picture of himself snaring a rebound. "That was me," he said.

"Without the hat," I said.

He gave me a look as if to say, There's only going to be one wise guy in our relationship, short though it might be, and it's not going to be you. I liked the hat, though. Heyman's hat gave him the mystique of a man who wanted to be noticed but not altogether known. It suggested a man who liked to see himself as a badass and preferred you to see him that way, too.

He kept staring at the picture of him up in the air, yanking that rebound. "Yeah. I was a great leaper," he said. "I could really jump."

It had been a few months since then. Heyman and I had agreed to meet at Tracy J's around four. Heyman showed up more than an hour late and immediately started screaming at Kristin, the manager who was tending bar. "Not even a soda!" he yelled. "He didn't even buy a soda! Not a beer or a soda! He's taking my time and he didn't even buy a single soda! These guys are all the same."

Pad and pencil in hand, I looked around the bar. Who was the "he" Heyman was shouting about? The room stood empty except for me and Kristen and Art. She looked concerned, mopping the bar top even though the wood already shone. It dawned on me more slowly than it should have that Art might have been referring to me—that is, I was sitting at his bar, drinkless, and he was shouting at Kristin about some guy who wouldn't even buy a soda. There were no other candidates for the position.

I didn't usually drink before an interview, but it now seemed that a drink was called for. "I'll have a Bud," I shouted, pounding the bar to show I meant business. "And how about I buy you whatever you're drinking?" It was Heyman's bar; shouldn't he be setting up the rounds? But who knew the etiquette of such moments?

"I don't drink!" Heyman yelled from the far end of the bar, with Kristen shouting in tandem, "He doesn't drink!"

"Well, how about a soda?" I ventured. "Let me buy you a soda." I was practically begging.

"I'll get my own soda," Heyman said, waving me off.

Then Heyman sat down at the bar, put on a pair of reading glasses, and began going through a sheaf of bills. I finished my first beer and started in on another that Kristen thought it advisable that I have. At last, Heyman pointed to a table in the corner near the window, and it was there in a quiet bar on a darkening afternoon that he told me the story of his life so far. I nursed a beer; he nursed a grudge—a lot of them, actually. But that was part of his charm, as it turned out. This was a man who knew how to hate. He still hated the North Carolina players—Larry Brown, Doug Moe—he'd tangled with back in 1961. He spoke about assholes with whom he'd battled the way other men spoke about their wives and children: fondly, with an appreciation for their particular foibles, for their petty meanness and shrunken souls. He appeared to exult in the world's chicanery. For one thing, that meant he recognized it. Nobody was putting anything over on Art Heyman.

By Heyman's senior year at Oceanside High School in 1959, the tracks of the so-called Underground Railroad between New York City and the North Carolina universities were glowing red-hot from all the traffic. Where once the Underground Railroad between North and South had served as a secret route for slaves escaping north to freedom, it had devolved nearly a century later into a subway train shipping Irish Catholic and Jewish hoop prodigies south from New York to basketball factories, where they would manufacture reputations for themselves and their universities. Within a generation or so, North Carolina would be able to stock its schools with local talent, such as Michael Jordan, James Worthy, David Thompson, and John Lucas, who'd grown up watching the first wave of black Northern imports such as Charlie Scott and Charlie Davis.

Heyman wanted to get on the train. He was a first-team high school All-American, six-foot-five and 205 pounds. As he puts it, he could shoot, rebound, pass, and jump. "I signed my scholarship papers first to North Carolina. Everybody down there was from the New York area. I was going to be one of those people because I knew Larry Brown [then a point guard at North Carolina from Long Beach, New York, now the coach of the New York Knicks], who was two years ahead of me. I had a hundred fifty scholarship offers.

"I had decent grades, so my parents wanted me to go to Stanford or Dartmouth. But I wanted to go big-time, and I picked North Carolina."

His visit to UNC proved memorable mainly because Heyman's stepfather and Frank McGuire, the North Carolina head coach, yet another New Yorker, nearly got into a fistfight at the Carolina Inn. Heyman had been out touring the campus in the company of North Carolina player Danny Lotz and came back to the hotel room just in time to witness a verbal brawl. "McGuire and my father were screaming at each other because my father called North Carolina a basketball factory. They were getting into it so I had to break up the fight."

But it wasn't the fight that made Heyman say goodbye to North Carolina. It was Vic Bubas. Later that spring, Hal Bradley, the Duke head coach, left to take over the Texas program, and Bubas, 31 years old, was named to replace him.

Bubas, who had played for Everett Case at North Carolina State and then served as his top assistant for a number of years, chased after top recruits across the nation with a ferocity that soon forced the other ACC coaches to match him. Refusing to concede New York City to Frank McGuire, Bubas targeted Heyman as his prime recruit as soon as he got the Duke job.

"On Tuesday afternoon, Bubas was named head coach," Heyman said, "and on Wednesday he was up here recruiting me. I'll tell all of Manhattan, he didn't make that big of an impression, because I was set to go to Carolina." Bubas did please Heyman's parents, though. They regarded Duke as academically superior to North Carolina, and that forced the issue. "I didn't want to go there," Heyman said. "My parents made me go there. They loved the school." Letters of intent did not bind a recruit in those days, and Heyman soon announced for Duke.

In Heyman's mind, however, his commitment was hardly a done deal. "Even when I was flying down to go to school, if Bubas hadn't picked me up at the airport, I think I would have gone over to Chapel Hill. But Bubas found out the flight I was on from my parents, and he got to the airport and he got me to Duke, boom, he put me in the room."

Heyman was a wickedly entertaining archivist of his basketball career. I appreciated the way he had polished his experiences into the smooth stepping-stones of story. He occasionally eyed me as he spun an anecdote, gauging my credulity, but it didn't particularly matter to me how precisely faithful to the facts his recitations were. I wasn't looking for an accountant's truth. Also, he didn't have that modest North Carolinian habit of keeping the meanness out of his tales. He relished revenge, he craved payback, and he exalted triumph—that is, if they were his own. He wouldn't have made much of a North Carolina Baptist or Presbyterian, but then he didn't have to be. He was a Jew from New York.

At Duke, Heyman went his own way. "I did things on my own terms," he said, "and I was like a legend there the first day. We were playing with the varsity guys and they were pushing me and I almost beat the hell out of them because of that. I made my reputation then. It was all over the campus. They wanted to see how tough I was, and they found out. I said to this one guy, 'Shove me again,' and he did and I threw the ball at him and whacked him and that was it."

Up in the stands that day, the freshman coach Bucky Waters had sat with several other coaches, watching the shenanigans. "They said they had an orgasm," Heyman said. "They said they finally had a tough guy to go with the other players."

His sterling play made an impression in some very odd quarters. His junior year, he received a telegram from Tuscaloosa, Alabama, from the Imperial Wizard of the Ku Klux Klan, Robert Shelton. "It said that as I go to a white Methodist school, I should uphold Christian supremacy.

Signed, the Imperial Wizard Shelton. He never thought I was a Jew. I was too tough, so he just couldn't believe it. He thought I was a Christian from up north. So he says, You should uphold Christian values, and good luck in the coming year. He was a Duke fan."

Heyman chortled at this and offered to dig up the telegram for me. "Jews were tailors and bankers," he said. "Nobody could believe I was this tough. I was bar mitzvahed and everything. Not that I was religious." This was clearly a theme for Heyman: tough Jew, tough among the anti-Semites, tough among the Southerners. An outcast wherever he went, this wandering Jew with a jump shot, but he would knock the Gentile snot out of you.

With Bubas at the helm, Duke in the early Sixties wedged itself into what for most of the previous decade had been a two-way rivalry between North Carolina and North Carolina State. The Blue Devils' games against the Tar Heels almost instantly became bloodfests in which each school's Northern mercenaries took turns trying to knock out the opposition, sometimes literally. With his big reputation, his cocky demeanor, and his temper, Heyman attracted more than his share of abuse from fans and players alike.

In a freshman game between Duke and North Carolina played at a high school gym in Siler City, Heyman was putting on a dominating performance against the Tar Heels' Dieter Krause, a player every bit as brutish as his name. By late in the second half, Heyman had scored nearly 40 points, and Krause could retaliate only by hocking a big gob of saliva at the man he was guarding. "I was killing him," Heyman said, "and he spit at me and I spit back at him. I was walking down the court and he cold-cocked me sideways. I was knocked out. Took stitches in my face."

Bucky Waters, then the Duke freshman coach, told me how he had warned Heyman about what was headed his way that week. "I told him, 'Art, you are going to feel heat in one way or the other because you jilted them,'" Waters said. "Frank McGuire is not one of those guys who turns the other cheek easily."

Heyman told Waters, "I know. I'll be okay."

Then the game started. "Every North Carolina guy that got near him called him every Jewish slur in the book," Waters said. "Christ killer, this and that. I'm watching Arthur, and his nostrils are distending and his pupils are dilating. So I call time-out. I'm twenty-three, but I'm pretty savvy. I got Arthur by the shirt. I said, 'Arthur, this is a test.' I said, 'If you're a great baseball hitter, they throw inside and knock you down. They brush you back. If you're a .180 hitter, they throw right down the middle.' I said, 'You're going to be a star. This is a test. If they can get you to lose it, they win, no matter what. He's bright. He goes, 'Oh.' I said, 'Arthur, I'll tell you what. You just kick their butt. If you play like you can, you're going to beat the snot out of them. But I'll give you one little retaliation. If they're still in your face, I'll let you point to the scoreboard when we start to dominate them.' He loved it, and he did. Then we spread the floor, and I should have taken him out. Bam! Down he goes. I had to take him to this little hospital in Siler City."

It was true; Frank McGuire had not looked charitably on Heyman's defection, and his players echoed their head coach's ire. Heyman said, "When I was visiting North Carolina, McGuire said to me, 'What schools are you going to visit? I said one more. He said, 'Where is that?' Duke. He said, 'I'll take you over there. I hate that school.' Everybody at Duke used to imitate him. They used to say, '*Thank you*' the way he did." Heyman pronounced a mincing thank you. "He talked like that."

"They had Cameron Crazies back then?" I asked.

"Oh, we started the Cameron Crazies," Heyman said. If the rivalry was a lightbulb, he was Thomas Alva Edison, its inventor. As he saw it, he showed up from Long Island and lit up the gyms throughout the region with the electricity of his presence.

"This was a civil war down there," Heyman said. "Big-time. And I started it. It was never that bad until I started it and it got nasty."

"What do you mean, you started it?"

"My betrayal of Carolina. Betrayal and the fight. Carolina had been the preeminent school, and then Duke became it."

The fight. One of the most famous battles in that civil war remained the game in 1961 between Duke and North Carolina at Cameron when a massive fracas broke out late in the second half. And Art Heyman, in his sophomore year, found himself right in the middle of it, a six-five, 205-pound light heavyweight firing punches at everyone in light blue. In his recollection, the brawl was triggered by a dollop of spittle, like the fight in Siler City his freshman year. (If Heyman was to be believed, there was more spitting between players in the early Sixties than in a catfight between Paris Hilton and Nicole Richie. The hardwood must have flowed with saliva.) "Doug Moe [a North Carolina forward and later coach of the Denver Nuggets] was spitting at me. So I said, 'The next time you do this, I'm going to fucking spit back at you.' And he spit at me and I spit back at him. And he walked up to me and I said to him, 'I'll fucking break your fucking head and kill you.' And that started it."

But the fight itself did not break out until near the end of the game, when Heyman hammered the North Carolina point guard Larry Brown, his old acquaintance from Long Island. "I fouled him and he threw the ball at me and I coldcocked him. Then Donnie Walsh [now the president of the Indiana Pacers] came in and I coldcocked him. And then a whole bunch of guys charged in. The fight was right in front of the Carolina bench, and Frank McGuire came up and kicked me and people said they then saw a hand come out and I hit him right in his nuts."

"You hit who?"

"McGuire. Right in his nuts. Years later, when I was playing pro ball, he went to a game and he came up to me and he smiled and in his Irish way, he says, 'Geez, it still hurts, Art.'"

In his three seasons on the varsity, Heyman averaged 25 points and 11 rebounds a game, but he could never quite take Duke to a national championship. In 1963, his last season, it looked like Duke, which by then had added Jeff Mullins, might go all the way, but the Devils ended up losing in the semifinals of the Final Four. Heyman boiled in the sour juices of his disappointment. He needed the release of an escapade, a road trip. "I was pissed," he said. "So I took a girl to Myrtle Beach, South Carolina, and I went into a motel with her. I signed us into the hotel as Mr. and Mrs. Oscar Robertson. But the guy at the motel recognized me. So they called the cops and they put me in jail."

"What was the deal?" I asked. "Was she underage?"

"No," Heyman said. "She was nineteen. But she was a student. Which means she should have been twenty-one." I wasn't quite sure what he meant, but clarification was hard to come by. Playing by the rules was not Heyman's strong suit. And anyway, he was rolling on this story.

"And the girl was crying," Heyman continued. "At the jail, they allowed me a call. I called Bubas. He said, 'Stay where you are. Don't open your mouth.' He calls up the governor of North Carolina, who was Terry Sanford. And Terry Sanford had a hotline with the governor of South Carolina. And the governor was home. So he called the lieutenant governor, who was from Myrtle Beach. In a half hour's time, the lieutenant governor came in and got me out. He put me in a South Carolina state trooper car and drove me back to Duke."

But that wasn't the end of the story. There was high-level perfidiousness to come. "Years later," he said, "when Terry Sanford was president of Duke in the Seventies, he decided to run for president. If Jimmy Carter hadn't been around in 1976, Sanford probably would have gotten it. So he calls all the Duke alumni, all the big-timers. So everybody was giving donations, and I wrote him a check for two hundred bucks. Now, Terry liked to have a few drinks. So, he calls me on the side and says, 'Art, we need to talk.' He leads me into his private library there and he rips up my check and he says, 'Make that for five hundred.' He says, 'Remember fucking Myrtle Beach! I got your fucking ass out of there.' And I gave him a check for five hundred. That's the only time I gave a donation to anything political. I said to myself, I hope this fucking check bounces."

When the NBA held its annual college draft that year, Heyman was selected number one, going to the New York Knicks. "I was rookie of the year," he said. "But I never lived up to my reputation with myself. I was kind of an outcast, you know. I was a white guy from a white school in a black man's game." He ended up playing in the ABA, from which he retired in 1970 after a seven-year professional career.

Like many former athletes, he went into the restaurant business. In New York, Heyman said, this line of work had necessitated a few close encounters with the Mafia. "Oh, it's big-time with the Mafia and me," he said. "They love me, and I went around and the FBI wanted me to become an informant and the IRS got on my ass because I wouldn't become an informant. And the fucking Mafia, they thought I *was* an informant."

Heyman might as well have been talking about facing a box and one, about a misguided opponent's attempt to take him out of the offense. North Carolina, the Mob, the FBI—he laughed at them all. His heart still beat with the old jukebox rhythms of dance-hall delinquency. He disdained the corporate, the buttoned-down and restrained. I remembered Heyman's phrase (referring to me) when he came in the bar earlier that afternoon: *those types.* His own coach, even, the vaunted Vic Bubas, fell into that category. "He was corporate," Heyman said. "You know, with the tie, the TV show, the camp. He could motivate, but he wasn't that great. I had a close relationship with him, but it was out of necessity, with all the trouble I had."

With enemies like that, who needed friends? It didn't matter who they were, Heyman took them on just like in the old summers on the Harlem and Brooklyn playgrounds. And then he would tell you all about the battles. There was not just vanity but generosity in his desire to entertain. He kept asking, "Is this good stuff?" He reminded me of the faux-gunslinger played by Warren Beatty in Robert Altman's film *McCabe & Mrs. Miller,* a sweet guy not nearly as tough as his reputation implied.

Just then, the phone rang. Heyman answered and in an aside to me, said, "This is why this place is successful. When I'm around, I do everything."

"That's exactly what Kristen said," I told him, kidding around.

"Everybody knows I'm an easy touch," he said, beaming.

The easy touch was unmarried now and childless. He had been close to his mother, but she had died a year and a half ago. He had his bar, which he'd named for his stepdaughter. He had a house on the Island and an apartment in Manhattan. And he had his history.

Duke had finally retired his jersey in 1990, 27 years after his last game. He attributed the delay to Duke's having retired Dick Groat's number in 1952. "They made a mistake with him," Heyman said. "People didn't realize he was thrown out of school for cheating. So Eddie Cameron got burned, and he said, I'm not getting burned again." Heyman credited his number being retired to Mike Krzyzewski. "He always treated me terrific," Heyman said.

He still kept a close watch on his former team and its chief foe. His views were typically pungent. He mentioned Michael Thompson, a six-eleven center who had transferred from Duke in 2003. "He wasn't going to play," Heyman said. "A big kid shouldn't go to Duke." This is exactly what Carolina fans had been saying for years: Duke University, where big men go to die. Erik Meek, Greg Newton, Chris Burgess, and now Shavlik Randolph—the list of highly touted post players who'd floundered at Duke was peculiarly long. And here was a Dukie voicing the same opinion.

He admired JJ Redick, who reminded Heyman of himself. "He seems like a tough kid. It's a funny thing—he was born the same day as me, June twenty-fourth." Clearly, Heyman had been reading that media guide as avidly as an 11-year-old boy looking for portents of greatness.

It was getting late. Heyman fished through a stack of mail and held out an invitation from Duke. It had a little picture of him on it. "A reunion or something. This Saturday. I'm supposed to go down there. But I'm not going to go."

"You're not going?"

"They think I'm going. Ah, I don't want to go. I'm estranged. I'm a loner, so when I finish, goodbye, good luck. I haven't been to Cameron in seven years. I lived down there for a while. When I sold some antiques out of my house—seven-thousand-square-foot house—the *Durham Morning Herald* ran a front-page story that said I needed money. And Duke, they didn't treat me well. I needed a ticket sometimes. They would give me a ticket in the last row down in Charlotte and that sort of shit. That's why I'm so bitter."

He said that when he needed tickets, he asked Dean Smith for help. "He's the best. I would have gone to North Carolina if Dean had been the coach. He's the nicest man."

A few customers were trickling in, hunkering down over the bar.

"You didn't think of me this way," he said, grinning. He appeared pleased at this possibility.

"I didn't," I said.

"I was a hero," he said.

When I shook his hand to say goodbye, he tipped his hat for the first time all night, giving me a fast look at a noble bald brow and a sweat-damp fringe of hair.

CRAZY TOWEL GUY

In the pantheon of Duke fans, there was probably no one who outstripped the ardor (and Carolina hatred) of Crazy Towel Guy. We'd met outside of Cameron Indoor Stadium a couple of weeks before on the night of the first Duke–Carolina regular-season game. Crazy was standing at the head of a long line of Duke students from Krzyzewskiville who were in the process of entering the gym in tent-sized groups after having their credentials checked by the head-of-the-line monitor. The students were being watched by fans and camera crews alike, a fact of which they seemed highly conscious. When camera lights would swoop over them like the search beam from a police helicopter, they would cluster together and shriek and shake their fists and waggle their fingers. They were screaming something that sounded like "ARGHHHOUEEEIIIO-UUUUU!" Or maybe it was "GRRRRRROOOOOEEEEEEEOOOO!" The Satanic language of the Collective Beast. And inevitably followed by, "Go to Hell, Carolina, Go to Hell!"

The screams and yells fed my anxiety. I was a spy behind enemy lines, and I didn't know how long my cover as a journalist would hold. I expected that one of the Crazies would see through my disguise, point a bony finger at me, and emit a high-pitched squeal right out of *Invasion of the Body Snatchers*. I would run into the scrub pines behind K's tower and conceal myself while the mob hunted for me, ready to tear me limb from limb and parade my deceitful Tar Heel remains right into Cameron Indoor Stadium for the pregame festivities. There, they would bodysurf my corpse up and down the stands the way they did the living corpse of Dick Vitale.

I noticed, however, that when the camera lights went out and the Crazies were cast back into the unhappy darkness of the unfilmed and unphotographed, they ceased to be a terrifying mob and reverted to being little clusters of gawky college kids waiting in line for a show. They chatted about courses and music and dates.

Then the lights shined on them again, and on cue, the monster was born anew. The Crazies raved and bared their teeth and cursed the Tar Heels and shook their fists and jounced their posters into the air. Later on, back in their dorms, they could watch their own highlights on ESPN, heroes of the new theater of self-involvement. They were following a script that by now was more than 20 years old: *Meet the Cameron Crazies, the Wackiest, Cleverest Fans in All of College Basketball*. Or, *The Real World: Durham*. Say it loud, I'm wack and I'm proud! Coach K had told them that the Crazies constituted the team's sixth man. Dick Vitale had told them that they were 1500-on-their-SATs nut jobs, baby! Blowing off a little steam before heading back to the library! They were crazy, man—at least, when the lights came on.

At the head of the mob, like a proud parent or a preacher beaming over his charges, stood a tall, lank-boned man with thinning hair and glasses, probably in his midsixties. Aesthetically, he was everything these kids were not: worn, angular, hollowed out a little by life. He looked like the grocery-store manager that it turned out he had been (though more precisely, he supervised a *lot* of grocery stores), the sort of decent fellow who might have ruled the cage to which the checkout girls scurried to get rolls of quarters.

He smiled beatifically at the Crazies and raised his hands. In one hand he held a white towel, which he began to whip in wild rings through the air, like a cowboy getting ready to lasso a steer.

En masse, the students began to chant. "CRAZY TOWEL GUY!" Clap-clap-clapclapclap. "CRAZY TOWEL GUY!" Clap-clap-clapclapclap. Two long claps and three speedy ones. This was the legendary Crazy Towel Guy. I had seen him from a distance at games, swirling the students into a beery froth of excitement. And here he was only feet away, engineering a well-practiced call and response that would have been the envy of any gospel preacher. Crazy Towel Guy seemed lit from within, radiating a luminescent fandom that made the other men standing nearby him appear old and timid, alumni reliving past lives, not men in the robust middle of their current ones. Or maybe the other men just seemed sane, sanity in such a context looking a little unimaginative.

I picked my way through the mob to Crazy Towel Guy and introduced myself amid the tumult and asked if we might speak when he wasn't in the middle of towel waving. It was a little like trying to have a normal English conversation with a man speaking in tongues. His towel just missed stinging my ear, but I don't think he intended that. He noticed me just enough to hand me his card. "Not now," he said. "Later." I headed out of the cheering scrum, followed by a friend of Crazy's, who told me that I should know that Herb Neubauer—Crazy's government name, as they say in Baltimore—was one of the "nicest people you'll ever meet."

So on a bright Saturday afternoon in late February, I set out for Crazy Towel Guy's house to watch him watch a Duke game. He lived on the edge of Durham in a development several miles out Guess Road. Driving there, I passed a string of churches, an establishment named The Hairport, and Priscilla's lingerie shop, which had a sign announcing that this was "where fantasy meets reality." If you were speaking of G-strings, as I suspected Priscilla might be, that meeting between fantasy and reality could certainly be arranged, but I suspected it might be painful.

But fandom increasingly struck me as the place where fantasy hid out from reality, like a gang of bank robbers hunkered down in a country shack. Fandom positioned you in an endless here and now of current seasons. The athletes always remained the same age, and if, by chance, you slipped off into reveries of past years and former players, you tended to remember them as they were. That such men went on to have lives and families outside of your purview seemed superfluous. Not to them, I imagined. But to the child inside the fan who is father to the man.

You had the luxury of ignoring their fumbling attempts at civilian life. Michael Jordan, for instance. How could he have had a second career that compared to his first, given that in his first he was only the best basketball player in the history of the universe? What could he do for an encore? Cure cancer? Or, even more amazing, get the Washington Wizards to win? Chances were that such a guy, a fellow addicted and accustomed to winning, might have trouble with what the novelist Walker

Percy calls re-entry. What does a god do when he is forced to retire from the universe he ruled?

Engaged in these melancholy reveries, I overshot Dover Ridge, Neubauer's subdivision, did a quick U, and entered yet another synthetic Southern neighborhood hacked out of the pine scrub in the borderlands where the city, in this case, Durham, began to dissolve into the country. The houses of the neighborhood were neat, unweathered, symmetrically arranged on boxy lawns. They were impervious to the locale, adherents to a national style that would have been familiar to dwellers on the nebulous edges of Scottsdale, Arizona; Washington, D.C.; or Columbus, Ohio. The houses looked a little stiff, as if they were strangers in town and didn't know exactly how to behave. They had tiny front porches that weren't big enough for sitting.

If you wanted the feel of home, you looked around these parts for old tobacco barns, those totems of the South that were fast disappearing from the landscape. In their weed-wrapped ruin, they signaled home in a way that actual homes had ceased to do. Not that you'd want to sleep in one.

Herb Neubauer was waiting for me at the door when I pulled up in front of his house. He'd just returned from taking his wife, Judith, to work at Belk's Department Store in Raleigh's Crabtree Valley Mall. Neubauer had invited me to sit with him and watch Duke play St. John's. This was extraordinarily gracious. Many obsessive fans preferred to catch the games by themselves, or at least with equally partisan friends. That way, they could scream, cry, howl, curse, moan, cringe, sulk, curl into the fetal position, and flick pretzels at the set, all without censure or regard for the feelings of anyone else.

I was lugging a paper bag stained promisingly with grease, holding two pounds of barbecue from Allen & Son, a barbecue joint by the railroad tracks on Highway 86 north of Chapel Hill. "Is that it?" Neubauer asked me. His clothes dangled off him. He looked hungry. I'd told him I was bringing the best barbecue he would ever taste for a pregame luncheon.

"If you don't like this, I'll know you're really crazy," I told him, bantering in the Southern way. It soon became clear that Herb Neubauer was something of a hybrid—not exactly Southern, and not exactly anything else either. But for all of his obsessive Carolina hating, he seemed about as sweet as a Moon Pie.

"I've got all kinds of soda," he said, opening the refrigerator to reveal various soft drinks. He didn't know of Allen & Son, but we stood at his kitchen counter and ate happily and fully: three plates apiece, slathered with slaw, and on the side we chipped in the occasional hush puppy. "I've had three heart attacks," Herb told me cheerfully as we attacked the barbecue.

"If you eat like this very often, you're going to have another," I said. "But it'll be worth it."

Neubauer laughed. "I've had three wives, too," he said with the jaunty tone of a natural-born survivor. Things clearly came in threes to Neubauer. Wife number three arrived from a little country village in the Philippines. They'd been together ten years. They had timed their honeymoon to take in a Duke football game against Florida State. They went back to her village each summer. On the eleventh of every month, he gave Judith a card to celebrate their anniversary, July 11, 1995. She was working so that the couple had health insurance; Neubauer would not become eligible for Medicare until October 2006.

"How does she feel about your passion for Duke?" I asked.

"She knows I'm obsessed," he said. "She still has a hard time accepting all of it. But she's learned to live with it."

She had developed her own passion for the game, he said, but it was a quiet one. "If we're watching a game on TV, sometimes she'll have to leave the room if it gets bad. I coach from the sidelines. She says they can't hear you. I still give my input. She knows what's going to happen when I start screaming."

I doubted there'd be much screaming today. In the aftermath of various scandals, St. John's was fielding one of its weaker teams. We took our seats in the den on the second floor, what his wife called Neubauer's "shit room," Herb in a blue leather chair, me on a blue-checked couch with a blue pillow. Arrayed around us were little Duke figurines, Devils and so on.

"I'm not a homer," Neubauer announced when Shelden Williams

walked on the first possession of the game. And this turned out to be true. He wanted Duke to win fair and square. The game was being played at Madison Square Garden, where Duke and St. John's had played some barn burners over the years.

Shortly thereafter, Lee Melchionni nailed a three, and Neubauer held his hand in the air. He wanted me to slap him five. "You have to do this," he said. "Every time someone hits a three, I have to do this with whoever I'm watching the game with. It's what I do." I looked at him, looked at his hand. I was torn between honesty and politeness. If only it weren't Melchionni, whose pumped-up manner on the court I detested. Every time he hit a three, he acted like he'd just single-handedly vanquished an enemy platoon.

"Come on," Neubauer insisted, his voice squeaking. "We've got to do this." He was begging. Somehow, I had neglected to make it sufficiently clear to Neubauer where my allegiance lay, even though I'd told him I'd grown up in Chapel Hill. On the other hand, I was a journalist, an anthropologist in my own backyard, and, most of all, an insatiable voyeur, wanting to see whether there was a difference between someone like Herb and someone like myself. So when in Durham, I concluded, do as the Dukies. Just this once, and don't tell anybody.

I whapped him solidly in the hand. "All right!" he said, his face lighting up. I understood this. I felt it. This could have been me. In most ways, it was me. Herb Neubauer could have been my mirror image, my alter ego. He merely bled a darker shade of blue.

But in other ways, Herb Neubauer fit the classic paradigm of the Dukie. He was born in East Orange, New Jersey. He attended Montclair Academy, a nearby private school. Then his father died and his mother moved the family back to her hometown of Rockingham, North Carolina. "We had no roots left up there," he said. Neubauer finished at public high school and won an academic scholarship to Duke in 1959.

"My mother didn't want me to go to a state university," he said. "She always felt we were better than that."

When he first arrived at Duke, he felt lost, ill-equipped for college. He believed himself to be at an academic disadvantage, coming from a

North Carolina high school. He'd enjoyed sports, but he wasn't good enough to play at Duke. He became manager of the Duke baseball team and befriended athletes at the university, among them Art Heyman.

"Art borrowed my car one time," Neubauer said, "while he was dating, I think, a girl from Chapel Hill. And he came into my room. I was still studying at 2:30 that night, and he said, 'I had a flat tire.' I said, 'Did you get it fixed?' He said no. 'I'll tell you where it is.' I said, 'What do you mean, you'll tell me where it is?' So we both went and found the car and got the car fixed, and I didn't really check it out very good. The next day I was riding from East Campus to West Campus, and girls would come down behind one of the dorms to wait for the bus and you'd just take them over to the other campus. So I picked up this girl and the girl looked on the dashboard and there was a used condom up there."

Neubauer had been in attendance at Cameron at the famous game in 1961 when the fight broke out between North Carolina and Duke. "Art took a lot of bad treatment and tough fouls in that game, and finally, Larry Brown gave him another shot and that was about all it took. The next thing I knew, it was bedlam, people rushing onto the court."

He tried to join the fray but fell through the bleachers and nearly broke his elbow.

After graduating from Duke in 1963, Neubauer bounced around the country for the next couple of decades, living in Charlotte; Washington, D.C.; Philadelphia; Denver; Charlotte again; Columbia, South Carolina; Newport News and Richmond, Virginia; and Durham, laboring for a variety of enterprises, among them First Union, Kinderfoto, Century 21, and, finally, the grocery chain Food Lion. In 1974, he sat in his Charlotte apartment complex on Monroe Road and watched Duke blow an eight-point lead to Carolina in 17 seconds. "It's the only time in my life that I destroyed the TV," he said. "I lived on the second floor, and when the game was over, I said, I'll never see Duke lose again on this TV. I threw it off the balcony. It was gone."

I didn't tell Herb that about a hundred miles away that afternoon, while he was chucking his TV into the parking lot, I was going crazy with joy. If you were a Carolina fan, that game spoke to the elasticity of time, the wisdom of fighting to the end, the occasional eruption of the miraculous into the world. That comeback seemed a reward for a morality that valued teamwork and process more than final results and yet paradoxically provided more than its share of glorious finishes. And all of that was available to any Carolina partisan merely by virtue of his emotional identity with the team.

If you weren't careful, those eight points in 17 seconds could puff you up with grandiosity like a Cheese Doodle. (Elasticity of time! Eruptions of the miraculous!) But my fellow Chapel Hillian Peter Cashwell made the best case for that near quarter-minute in an essay entitled "Seventeen Things I Learned from Dean Smith," which alluded to *The Great Gatsby*. "I became an optimist over a 17-second span." Cashwell writes. "The green light may wink across the bay, but experience teaches us that it will almost certainly remain out of reach. Almost. But I cling to that "almost" as if it were a lifeline, and I cling because of that impossible 17-second span." So profound had been the effect of that game on a generation of UNC fans that it had produced its own poets, a rhapsodic literature of the miraculous. That old Ralph Waldo Emerson, coach of the New England Transcendentalists, could have done no better.

By contrast, the St. John's game we were watching was lackluster. Five minutes in, St. John's led Duke 6 to 3. Neubauer was irked by Shavlik Randolph. "He's a pussy," he said with the pure venom that only a fan whose expectations of greatness have been thoroughly dashed can muster. "I blame his parents. The way they raised him. When he got to Duke, he didn't even drive a car!"

With 13:25 left in the half, the score was only 8 to 5. The game was ugly. Redick already had two fouls and was sitting on the bench. "Come on, Mike!" Neubauer screamed at Krzyzewski after the Blue Devils turned the ball over once again. "Mike says there's not a leader on this team. Josh McRoberts and Greg Paulus [incoming freshmen for the next season] are leaders. Mike says they'll start.

"I screamed at the TV last week," Neubauer informed me, annotating his outbursts. "This is a tough game to watch."

At 11 minutes to play in the first half, Duke had hit two of 14 from the field. On cue, Melchionni botched a layup. "How'd he miss that chippy?" Neubauer squealed. At age 64, he had a voice that still rose and cracked like a teenager's when he got excited. And seconds later, he became excited again when Shav had a rebound poked away from him by a St. John's player. "That's a pussy!" he screamed. "*See!*"

At this point, Duke was three for 20 from the field, according to the announcers. "Three for twenty," Neubauer sighed. "*Sheesh.* My wife would have gone into the other room by now." He was becalmed, trying to accept his fate, which was the fate of the Blue Devils, who this afternoon seemed laggardly and uninspired, like a team that might not have it. Neubauer had to conserve his energy for later in the contest.

When DeMarcus Nelson dropped in one of two free throws, Duke finally tied the game at 11. "Can you believe this?" Neubauer cried. "Eleven points."

Ordinarily, I would have been rooting like a madman for St. John's. The Blue Devils were ripe for a fall this afternoon, that much was clear. But it was more than politeness that constrained me. No, I was starting to realize that Neubauer was my doppelganger, that he was allowing me to peek at him in the privacy of his own rituals and hopes and fears. There was such innocence in Neubauer. He cared this much about a game. That made him like a child.

So when the Duke freshman David McClure knocked down a three and Neubauer put up his hand to be slapped, this time I didn't hesitate. I stung his palm hard, and to tell you the truth, it felt good. The slap reverberated through the den. It occurred to me that despite being rabid partisans of two diametrically opposed teams, Herb and I probably had more in common with each other than we did with the athletes we were cheering.

"Mike is going to singe some asses at halftime," he said with glee.

"The referees already look a little singed," I said.

During a commercial, Neubauer started talking about Luol Deng, who'd starred as a freshman at Duke the previous season before turning pro a year or so earlier than expected. "K was angry about Deng and his family," Neubauer confided. "Mike said, 'Fuck him.' Deng had told him he was going to be back. Of course, that was before Deng found out how many brothers and sisters he had. Fifty-seven!"

"Are you close to Krzyzewski?" I asked Neubauer.

"We talk," he said. "I send him a card every year on his birthday, and he always writes back. He's a really funny guy. I'm friends with his brother, Bill. After we beat Nevada–Las Vegas in the 1991 Final Four semis, we were drinking in his hotel room late one night. Mike was watching tape in his. Around three A.M., he poked his head in from his adjoining room to see what the hilarity was all about. 'Mike, you're still watching tape and we're still drinking,' his brother said. 'You're doing your thing and we're doing ours.'"

Shav went up weak with an inside shot. Neubauer slammed the arm of his chair. "High school!" he screeched. "Go up with the fucking ball!"

Melchionni eased the pain by nailing another three. He alone of the Blue Devils was enjoying a good game. Neubauer put up his hand. And again, I slapped it. By the half, Duke had taken an 11-point lead, 28 to 17.

At the half, Neubauer took me into the spare bedroom where he kept mementos of his decade as Crazy Towel Guy. In 1994, his apartment had burned down, destroying his memorabilia: championship balls and tickets and pictures and over twelve thousand dollars' worth of framing. Stored here were the souvenirs he'd accumulated over the last ten years: a framed towel with the legend "The official towel of the Cameron Crazies," a *Newsweek* article entitled "Ode to the All-important Sixth Man." He laid a bundle of items on the guest bed.

"Look at this," he said, proudly showing me a newspaper clipping from the year before, ranking him number 12 on the list of most memorable characters in ACC history, compiled during the conference's fiftieth season. And there he was, with such legendary figures as the coaches Lefty Driesell, Bones McKinney, Jim Valvano, Norm Sloan, the referee Lou Bello, and the North Carolina radio play-by-play man Bill Currie, known as "The Mouth of the South."

It seemed as logical a time as any to ask Neubauer about the birth of Crazy Towel Guy, a story that began, as it turned out, with a blue leather hat purchased in 1984 on a side trip to Tijuana, Mexico, during the weekend Duke was playing in Los Angeles at the Bud Light Trojan Classic.

"I saw the hat," Neubauer said, "and I thought, *Wow, that looks just like a Duke hat. I'm going to wear this hat to every Duke game from now on.* And I did. I sweated all over the country because of that hat. People would throw things at it. At NC State, they used to hit it with gum. At St. Johns, people would yell, 'Pancho!' at me. At Georgia Tech one year, a Tech fan grabbed it and threw it in the middle of the court like a Frisbee. I sweated so hard in it that I started taking a towel to the games, to wipe my head."

Then, in 1994, when his apartment in Durham burned down, he had only enough time to attempt to wheel the TV out the door and grab the letters Judith had been writing him from abroad during the pen-pal part of their relationship. The hat went up in flames. This was devastating to Neubauer, but, well, there were still towels.

"So," he said, "I continued to take the towel to Duke games," he said. "Because, you know, I continued to sweat, even without the hat. Cameron was hot and I'd get excited, and every time Duke scored, I'd wave my towel. And eventually, I noticed that the students were yelling, 'Crazy Towel Guy.' So I'd get up and wave the towel for them and it got to be a habit."

A hat was lost and a legend was born. He became "that crazy towel guy," and soon, in the American way, he had a brand. He was "Crazy Towel Guy," the fan as superhero. He appeared on TV, got written up in the newspaper. With his fame, Neubauer took advantage of his brand identity and cooked up a few business schemes. He handed me a bottle of Crazy Towel Guy water. On the label was a picture of skinny Herb and the towel. There were more bottles in the garage, he told me. "I think this'll probably be enough for the time being," I said.

Neubauer said that he generally used two towels a year, employing the same towel until Duke lost a home game. He then washed that towel—he laundered them himself—and began to use the second. Some years, he donated the used towels for auction by local charities.

The way Herb figured it (and he'd done the math), since 1997 the towels had a winning percentage of 94.7 overall, and 87.5 against UNC.

The second half began with a close-up of Krzyzewski visibly exclaiming, "Shit!" after Daniel Ewing lost the ball. "He's mad at Daniel," Neubauer said. "That was a lazy pass. Mike won't be happy."

Neubauer knew how Krzyzewski could slice you up with his tongue. He and a friend had once asked the coach why he hadn't played Corey Maggette more in the championship game loss to Connecticut in 1999. "How many championships have you guys won?" Krzyzewski answered.

Duke's 11-point lead gradually melted. "Come on, fellas," he pleaded. "They need me to be up there! The crowd is terrible!"

With 7:48 to play, St. John's banged in a close-range shot that shrank the Duke lead to four, at 41 to 37. "I could have scored on that play!" Neubauer squeaked indignantly. "Shav jumped before the guy put it up!" He slapped his knees and exhaled. "*Shewwwwwwww*," he said.

I was with him in spirit, but I was also beginning to scent an upset. I knew the jujitsu fan move of speaking well of the opposing team when they were playing poorly, suggesting that they would come back, that they were too good not to win. To do otherwise, to announce the loss ahead of time, was to jinx the underdog and curse the upset. So I said, "Herb, relax, man. As bad as they're playing, the Devils aren't going to lose this one."

He moaned long and hard.

And then, defying my attempt to influence the game, Redick popped in a long three, putting Duke ahead by seven. He was now one of six from long range. "You get away from the ACC and you think you can relax, you think you can take a break," Neubauer said. "But you can't."

And he was right. Redick proceeded to miss another three, Shelden Williams picked up his fourth foul, and suddenly, with 6:47 left in the game, Duke clung to a precarious five-point lead, 44 to 39. I noticed Neubauer was starting to tell the Johnnies what to do, how to score. I suspected that he was using his own jujitsu. He wanted to preserve the Blue Devils by coaching their opponents to success. If your boys were going to lose, better to take an active hand in it than sit by passively, enduring one bad pass after another. Duke had thrown the ball away 21 times!

Just then, Redick air-balled a jumper. "Don't cut your hair!" he screamed at JJ, who had emerged for this game with a tight crop. But that was the low point. The game turned from there. Duke began to push the lead out a little further, and despite Neubauer's announcement that "you can't feel comfortable with this group of no-good ball-handlers," the Devils eked out—and "eked" was the word—an 11-point victory, 58 to 47.

"*Shewwwwwww,*" he exhaled again.

"Congratulations," I said, slapping his hand.

After the game, Neubauer and I lingered in our blue seats. The February light dimmed to evening. The house was silent. His wife was still at work. The Virginia Tech–NC State game flickered noiselessly on the TV. We could have been two superannuated veterans of the Civil War, one Confederate, one Union. Two once-bitter rivals who had more in common with each other than with the strange world that had gradually come on around them, all those young people who knew and cared nothing of what it had been like at Sharp's Ridge or Antietam.

Neubauer had suffered the first of his heart attacks in 1982, while working in management for Food Lion in Newport News, Virginia. So he moved back to Durham and stayed with the company as a supervisor until 1987, when at the age of 46, he woke up one day and realized what his purpose in life was, how he wanted to spend the rest of his days. "I decided that I'd be a full-time Duke fan," he said. And with that, he retired to follow his dream, as the expression goes.

His decision struck me as both brave and quixotic. His was the life of the spectator. Was that not a tad too passive, with the successes and the failures too connected to the collective efforts of dozens of others? But then again, how different was that from devoting your life to collecting seventeenth-century Persian miniatures, painted by the long dead, for instance? There was always the possibility that narrowing into the thing one loved ultimately provided the greatest possible opening.

"What is it about Carolina," I asked, "that brings out the beast in Duke fans?"

"It's the arrogance of the Carolina fans," he said.

"They would say the same about Duke."

"Whatever I might say," Neubauer said, "they're going to say the same thing. Certainly, they won the first national championship so they felt like they had lifetime bragging rights."

"Is the dislike for Carolina a good-natured thing for you? Or do you feel genuinely passionate hatred?"

"I think there is genuine hatred. When Duke beats Carolina, my buddies and I say it's a double victory. ADV for Another Duke Victory. And ACL for Another Carolina Loss. When we toast, we clang glasses twice. I have one Carolina friend, and we don't touch the subject very much."

Neubauer had slowly come to the conclusion that perhaps the two schools did embody different values. "Carolina people think because Duke's a private institution that Duke people think they're better than everyone. Now, going to Duke, I never really felt like that, but Duke sort of put that feeling in me. So they're right. I am better than them. They're finally right after all these years."

I asked Neubauer, "Is watching basketball a religious experience for you?"

"The intensity of it is. There were years when I was younger when it probably got the best of me. I couldn't sleep if Duke lost. Now I can go to bed. I can sleep. I realize there's going to be another day, another game."

"What gave you that equanimity?"

"Probably when I kept thinking that I almost died a few times. You get to the point where you say, Hey, I might not see a game where I'm going."

"Who knows?" I said. "They might have ESPN there."

"Depends on where I'm headed," Neubauer said with a grin. "Of course, there are other things in life than basketball. But this is what I really enjoy. My friends and I talk about how so-and-so would have enjoyed this or that game. But he's gone. As you get older, you never know how long you'll be here. The game makes me feel young. Gives me vim and vigor. And when I don't have that, I hope I'm not here to see it anymore. This is the golden age of Duke basketball. It's been a great ride. I don't want to forget it."

It was time for Neubauer to pick up Judith. He escorted me out to the lawn. It was one of those soft February nights that occasionally arrive in the Piedmont: warm breeze, scent of grass, intimations of spring. Winter would be back, but the air gentled you, played with your expectations. Neubauer and I slapped hands for the last time. A satisfying playground *thwock*. My palm was still stinging as I took the wheel. Just don't tell anyone.

A SPY IN THE HOUSE OF HATE

I started the drive home suffused with fellow feeling. I had driven the road between Durham and home a thousand times, but now something had subtly shifted. How would I explain this to my family? My mother had always tolerated my escapades, but this one might be harder for her to bear. And as for my sister . . .

The previous year, our neighbor Ted had given me two tickets to a late February game at Cameron, and my sister, Annie, had come along for the ride. After what happened at the Duke versus Valparaiso game, I'm no longer sure that it is fair for Annie to think of herself as shy. We were sitting in the Iron Duke section, good seats not far from the court. Duke rampaged to a 27-to-8 lead. The Blue Devils were bombing threes— Shavlik Randolph, Daniel Ewing, Chris Duhon, even Shelden Williams. A Valparaiso player scored on a layup and even that drew the Crazies' censure. "You can't dunk, you can't dunk," they sang.

Alone among the thousands, my sister clapped her hands on those occasions when Valparaiso scored or made a good play, which so far had been infrequent enough to keep us inconspicuous. But then she couldn't help herself. Shelden Williams lowered his shoulder into a Valparaiso defender and sent him crashing to the floor. No foul. "Can you believe that?" Annie asked, rising out of her seat to the puzzlement of those partisans around us. "I mean, really, did you see that?" she asked, gesturing at the floor, looking around for support.

"I saw it," I said.

She was my sister, and it was me and her against the world. Just like in the days when she used to hold my hand and accompany me upstairs, where I believed that Alfred E. Neuman from *Mad* magazine lived in a chest of drawers on the second floor and was liable to pop out and harass me when I was alone. She was younger than I was, but she was fearless. I felt no shame in dragging her up the stairs onto that dreaded second floor.

Everyone in our section was watching us out of the corner of their eyes, the way riders on the subway stealthily eyeballed potential trouble-makers.

The onslaught continued. Ewing hit another three, then Redick canned one and got fouled at the same time. It was 39 to 13. A Valpo player went to the line for free throws. The Crazies whispered, "Sssssshhhhhh," like a little wind sluicing through the stadium. Then, just as the shooter was about to release the ball, they screamed in unison. "That's really child-ish," my sister said, standing up again. "Childish! They are so childish!"

"Yes, they are," I said.

Valparaiso made a charge, such as it was, and narrowed the lead to 19. Just in front of us, a white-haired guy who looked like he'd done a mile or two got out a trumpet and feebly squawked a few notes. "Charge," responded a few sympathetic souls in the crowd.

Coach K jawed at a ref for calling a foul on Duhon. Was it a coinci-dence that on the next play a Valpo player was whistled for traveling? Duke's lead had fallen to a mere 16 points, at 43 to 27. Anxiety pervaded Cameron. But then Duhon hit a three as the shot clock expired. The Devils didn't panic; in that, they reminded me of the old Dean Smith–coached Carolina teams. Was there a lesson here about how all things must change, how no empire lasts forever?

The old trumpet guy certainly seemed on the verge of expiring. He wheezed and sputtered a few notes that may have once again signified "Charge!" It was hard to tell. The blood drained from his face, and his lips were turning Duke blue.

"Look at that guy," my sister said. "Should we call an ambulance?"

A few fans with exquisite hearing turned toward the trumpeter and thrust their fists in the air, yelling "Charge!" again. It was a kind thing to do. He sagged into the bleachers, white as a sheet, his head drooping onto his chest. He had done his part for the home team. Now he could die in peace. Don't tell me that they had no hearts, these Crazies.

My sister wasn't having any of it. "That's pathetic," she said. "He needs an oxygen tank just for that trumpet."

At the end of the first half, it was 53 to 29 in Duke's favor, and the second half brought more of the same. At one point, the band pulled their blue-and-white rugby shirts over their heads and started acting like monkeys. "Oh, *that's* funny," my sister said. With 10:57 to go, Luol Deng held off in order to snag a rebound. No call. He put up a layup, and at last the ref called a foul—on the Valpo defender. My sister went berserk. "That was a foul on Deng," she cried. She looked around her at the fans. "You know that was a foul," she said. "Is Coach K paying the refs?" I stood up with my sister to demonstrate solidarity. It was just like the old days, going up the stairs together into the realm of Alfred E. Neuman.

At 7:16, people were actually leaving early. But we stayed to monitor the proceedings, like observers from the United Nations. Nick Horvath was tagged for a foul on a rebound when he pushed the Valpo big man down. "That was mean!" Annie yelled.

At 2:42, she shouted, "Look at the Rat, still working the refs." I had to hand it to Annie; we'd managed to clear our section of Duke fans, a substantial accomplishment. The Valparaiso players hustled to the end, but Duke finally won, 97 to 63. Joe Paglinea had come in for Sean Dockery, which must have been the official bench confirmation of a blowout. "At least they didn't get to one hundred," Annie said. As we picked our way through the bleachers, the trumpet man was squeezing out a deathly version of "Hey, Hey, Goodbye." It sounded like the bleating of sheep being slaughtered. "He should save his breath for staying alive," my sister said.

Fans backed away from us. We enjoyed a clear path to the exit. It was as if we were carrying weapons, which in a sense we were: We had my fearless sister. "They're such wussies," she said of the Crazies. "I mean, really, have you ever seen such wimps?" My pride knew no bounds.

KRZYZEWSKIVILLE

One afternoon during his mission behind enemy lines, the journalist strolled among the pathways of K'Ville. Much visited by reporters like himself, the place had the look of a squalid refugee camp for the children of the upper classes. Here were sagging tents pitched in a muddy yard behind Cameron Indoor Stadium; here were their occupants clad in polar fleece and jester's caps, checking their e-mail on laptops tapped into nearby wireless transmitters.

In the open air, they were reading Shakespeare, physics, Jacques Derrida, molecular biology. He came across a young man named Omar poring over a Bible. "Is that for a course?" the journalist asked. "Uh, no," Omar said. "Just something that I am reading."

They were there, all of them, for the incalculable privilege of being the first admitted into the North Carolina game. Many had been tenting, as it was known, since New Year's week. Their faces were ruddy from life in the open air.

Three young men, bundled against the elements, invited the journalist into their tent, where they lounged on the floor. They asked that their names not be revealed, so that their parents wouldn't discover that the more than $40,000 a year they were investing in a Duke education was securing for their children a level of comfort more traditionally associated with the victims of monsoon or earthquake. "There've been studies at Duke," one said, "that show that the average freshman guy's GPA goes down .5 the second semester of his freshman year, what with tenting and fraternity rush."

They came from moneyed suburbs: Vienna and McLean, Virginia; Westchester County, New York; Evanston, Illinois. They had heard stories from earlier settlers about how back in 2003, the winter had been a hard one—snow, ice storms. Just like in pioneer days. Many K'Villers had gotten sick. Shivering in their tents, flu, hacking coughs, a medical disaster. Some had lived to tell the tale. So now there was a grace offered when the temperature dipped below freezing. They were allowed to go back to their dorms, though some toughed it out. "We're all proud of each other for sleeping out here," McLean told me.

The freshmen spoke reverently of Mike Krzyzewski. "He is *our* coach," Westchester said. They loved Dick Vitale, too, and looked forward to bodysurfing him over their heads, posing with him for cell-phone pictures that they would instantly transmit back home. When they considered their counterparts at North Carolina, they spoke with a fair degree of insight. "They think we're a bunch of rich kids," Evanston said. "They think that we all drive Beemers. They think that we've all come down into their territory and that we're going to kick their asses."

"We are," said McLean, laughing.

They were friendly, these students, even when the journalist confessed to having attended their rival school, and even when he admitted that he fervently wished for Duke to lose later that week. They laughed like comrades. He suspected that his age protected him—why bother this old guy skulking around the encampment? He was harmless enough. Hell, they were harmless enough. The previous spring, the Crazies had indulged in what had been described as the world's largest pillow fight the week of the Carolina game. It might also have been described as the world's largest act of mass weeniedom.

These tenters were actually interested in the journalist's career. How had he come to be sharing a tent with them, all hunkered inside like Bedouins at parlay?

"Cool," one said when the journalist described his career trajectory, though "trajectory" might have been too kind a word, since the actual arc had more in common with a flat line. The guiding principle behind his "career" had been serendipity. The students' curiosity should have come as no surprise, he later thought. The media, of which he was a tiny part, intrigued them. And there was ambition in this place—he could smell it above the wet wool and the faint scent of beer—and he was an example of the curious ways one's life might turn out.

By contrast, the students whom he had met at North Carolina struck him as less worldly. They hadn't peered with such intensity into the future to discern the shape their careers might take. They weren't plotting with the same precision that these youngsters were.

And yet . . . at every stop of his Hate Duke World Tour, the journalist had been searching for justification, for reasons to keep on hatin' on, to heap the rhetoric like burning oil onto the heads of the students here. But they were screwing up his plan. Two of the three freshmen—two of them!—wanted to manufacture prosthetic limbs. What were the chances of that? They might end up as lawyers or brokers, but right now they were telling reporters and girls that they wanted to help amputees.

Maybe the journalist was a wimp without the courage of his convictions—certainly a possibility, for it is harder to hate an individual than a group, unless perhaps you're in a "relationship" with an individual, in which case all bets are off. He certainly had not displayed his inner beast among these well-scrubbed, happy Dukies. Or maybe the more detestable elements of the typical Dukie were concealed like the headlights of a fancy sports car. Maybe underneath, these guys still believed in success at any cost.

Or maybe the cost was just $40,000 a year. Still a lot to pay.

A VISIT WITH THE RAT

All season long, I had been paddling up the Nile of my Duke hatred, looking for its source. And with every new meeting, my hatred evanesced like fog in a bright sun. The Duke players, past and present, even the Duke fans—I was mystified to say that I kind of liked them. They weren't that bad. Some were even better than that. Now, however, I was faced with the ultimate encounter, which I had long been dreading during my voyage upriver: I was going to see Mike Krzyzewski. Or, as he was more commonly known to me and my fellow Carolina fans, The Rat. Ratface. Fuckhead. Satan. The Evil One.

I would finally see whether Krzyzewski up close was the same guy I had hated so lustily from afar, whether man and symbol matched, whether there was any similarity between the dark prince of my psyche and the coach from eight miles down the road. A friend of mine mocked me. "You're going to like him," he said.

"You don't know that!" I cried.

"You're going to like him. *Nah nah nah.*"

The journalist in me was wary, because he understood one of the basic perversities of journalism: that at first encounter, people are rarely as bad as you might expect. Even the most cold-blooded of dictators exert a simple human charm, unless they're executing opponents in front of you, which they are usually polite enough not to do. Paradoxically, the more evil their reputation, the more striking their simple decencies appear.

Journalists love this sort of thing. By such paradoxes are stories generated. By such ironies are the easy structural moves of reversal accomplished. The old "on the one hand, but on the other . . . ," or the classic "he did this but he also did that." The killer in his cell was so gracious, so mannerly, so calm. It was hard to believe that he had eaten an entire family of Icelandic farmers.

When it came to Duke coaches over the years, I hadn't always been a raving maniac. I knew that I hadn't hated Vic Bubas (I was too young), Bucky Waters (his crew cut was so out-of-date that he inspired sympathy), Neill McGeachy (10 and 16), or even Bill Foster (don't know why not). In contrast, I couldn't remember a time when I hadn't hated Mike Krzyzewski.

If I didn't arrive today at the profane and stinking springs that fed my hatred, I feared I would never get there. And that by itself was a terrifying prospect.

So as you can imagine, it was all a little surreal and anticlimactic as he came towards me on an ordinary Thursday afternoon in the grimy bowels of Cameron Indoor Stadium. Here was a man I had cursed most foully for a quarter-century. "Sit down, Rat!" I had commanded just the other night. And here was a coach who had most foully cursed everyone around him for the last 25 years, who I imagined would have cursed me, too, had he only had the opportunity.

We shook hands firmly. He led me into his old office at Cameron, the one he inhabited before moving into the tower next door. This was a windowless bunker of a place, crowded with basketballs and Duke paraphernalia.

He was fidgety, sitting in his shorts and sports shirt and sneakers at an angle to his old desk in that old office, his legs stretched out in front of him. He jiggled them as we talked. His bony knees gave him an odd aspect of vulnerability, a kid at summer camp. But the intensity!

The famous Krzyzewski intensity radiated off him like the magnetic disturbance from a sun flare. He wore his game face. Before long, though, I would actually make him cry. Yes, I would.

But first, I asked him whether after 25 years in Durham, he considered himself a North Carolinian. He did, he said. He and his wife loved North Carolina. They intended to retire here. He loved the beauty of the place, and he loved to cultivate its beauty in his garden. "I don't like to golf," he explained. "I understand why people love golf. But I'm competitive in everything I do and then I would have to be competitive in golf. And I just want to be outside, growing things."

Outside, in his garden, Krzyzewski was different. He was like the immigrant gardeners from the Old World whom I saw in Queens, the wizened retirees from Italy and Greece and Malta, tending their little patch of dirt in their shorts and sandals and white socks, weeding, digging in the earth, cultivating, watering, raising figs and grapes and cherry tomatoes and eggplants. Carrying them inside to their wives as proudly as roosters.

About his garden, Mike Krzyzewski spoke this afternoon like a nature poet, a man in love with trees and shrubs and flowers whose names he was slowly learning. A Chicago boy with more nature in his grasp than he ever imagined existed. "A tree, a tree would be good here—that's what I think. I can see it, I can watch it grow," he said. "I don't know if it's being Polish or whatever, but when I'm outside I want to work. I want to cut something down, to prune a limb, to grow something." Even in the garden, he competed, he admitted. Against the deer and the insects, against too much heat, not enough rain. "Nature, I guess," he said. But not against people the way he did everywhere else.

In the garden, Krzyzewski got to be alone. Everywhere else he had to be Coach K, basketball guru, elder statesman, lightning rod, loved and detested in unequal measure. Some places he couldn't even go. North Carolina might be his home state but he only went to Chapel Hill once a year—when Duke played Carolina. He had never been on Franklin Street his entire life. "What would I do there?" he asked with a laugh. "Except cause trouble." But in the garden, it was different. "My fingernails get dirt under them," he said. "Feeling dirt. I love that. You know what I mean? I don't know whether it's good or bad, but it's what I like."

In his garden, he could be Mike Krzyzewski, just another Polish guy, an accidental, patriotic American, late in middle age, trying to make things grow. He'd traveled a long ways to this green place.

Born in 1947, Krzyzewski came from Chicago, like the hero of Saul Bellow's *The Adventures of Augie March*, and his own story seemed to echo that great novel's opening, the pugnacious, hungry, prideful announcement of a guy from the neighborhood: "I am an American, Chicago born—Chicago, that somber city—and go at things as I have taught myself, free-style, and will make the record in my own way: first to knock, first admitted; sometimes an innocent knock, sometimes a not so innocent."

Krzyzewski's parents, Bill and Emily, raised their two sons, Mike and Bill, now a retired fireman, in the first floor of a rented house on Cortez Street in a predominantly Polish neighborhood on the North Side. Relatives lived upstairs.

His father worked as an elevator operator for nearly 25 years, a quarter of a century spent inside a windowless cab, opening and closing doors, riding up and down, up and down. Good morning, sir, good afternoon, sir, good night, sir. He saved a little money and opened a lunch joint in one of Chicago's industrial sections. It failed. So then Bill Krzyzewski bought and ran a local bar on the South Side, which is what he was doing when he died of a cerebral hemorrhage in 1969, his son Mike's last year at Army. Mike hadn't realized the full extent of his father's illness; his family hadn't wanted to bother him.

That Krzyzewski had ended up at West Point playing for Bobby Knight, his first coaching mentor, he credited to his parents. Had he had his way, he would have gone off instead to a Big Ten school or Creighton University in Nebraska. His last two seasons of high school, he'd been the leading scorer in the Chicago Catholic league. His grades were good. Then Bobby Knight came calling. He wanted Krzyzewski to play for him at West Point.

No, thank you, Krzyzewski said. For the next week, his parents muttered to each other in Polish. "Stupid Mike," they muttered. They muttered within their son's earshot: "Stupid Mike, stupid Mike." Finally, Stupid Mike gave up. Resistance was futile. "When you have one parent

who never went to high school," he said, "and another who never finished high school, they looked at West Point as an opportunity for their
son to do something that only the rich and privileged got to do. I could
have let me down, but I couldn't let them down."

The world may not have been exactly rigged against the working
class, but it sure wasn't inviting it up to the mansion, either. Krzyzewski
came up believing that he was going to have to batter his way into a
society that would have been quite happy for him to pound the North
Side pavement for the rest of his life. And if Krzyzewski himself wasn't
exactly a working-class hero, he harbored no illusions about the way
the world worked. "It isn't all equal," he said. "If it were, everybody
would be born into the same pool and given the same advantages from
the start."

His father paid for the unfairness with his name. "He didn't go by the
name Krzyzewski," the son said. "He went by the name Kross. That's
how my father handled ethnic discrimination. That's what he felt he had
to do to get a job."

"It made me better prepared to handle this environment," Krzyzewski
said. "C'mon, what are you going to do to me that wasn't done to my
dad at a worse level? Because they were Polish, my parents were worried
that if they messed up in any way, they might not get the job. That's the
real world. So if you're going to say something bad to me about Duke, is
that going to hurt me? Are you kidding me? It's only going to make me
want to do better."

The son, it was becoming clear, would make up for his father having
to counterfeit the family name, an act of concealment that bespoke fear
and the cold sweat of ruin. Sportswriters everywhere would not only
remember the family name—they would have to spell it, too! That's
Krzyzewski! Take that, you journalists!

If Krzyzewski was any indication, the rocket fuel of American achievement was resentment, the desire to show someone, someone who never
noticed you or your family, never believed in you or your family or anyone from your neighborhood, that you were not to be ignored, not then,
not now, not ever. The past could never quite be shirked and the future
that would cauterize the stings of that past could never get quite close

enough. More wins, more wins, more wins. Shovel them into the furnace, the engine, keep that thing firing, keep it burning red-hot.

Three national championships, in 1991, 1992, and 2001—not enough.
Five straight ACC tournament championships, ending with last year's
loss to Maryland—not enough. At Duke, Krzyzewski had managed to go
as far afield from the South Side of Chicago as he could go, while at the
same time recreating for himself the sense of being embattled on the
block, embedded in hostile territory, surrounded. He was a Catholic in
the South, and there hadn't been as many of those when he arrived to
take over Duke in 1980 as there were now. It hadn't been all that long
ago that North Carolina had been considered missionary territory by the
Catholic church. But that wasn't the main thing.

The main thing was being at Duke. "With North Carolina and North
Carolina State around, we're a minority," he said. "But growing up in
Chicago prepared me for that. It wasn't so much race there—black-white—
it was nationality. I was Polish and I was a minority. And I think that's one
of the reasons I've done so well here. Because we're *still* a minority."

"You have to have a thick skin," Krzyzewski said. "If we did at a state
school what we have done at Duke . . ." He stared at me. "Pick a state,"
he said.

"North Carolina," I said.

"Not *North Carolina*," he said. He was trying to make a point about
North Carolina, I was to understand. In this state, Duke would never be
considered the home team.

"Okay. How about Rhode Island?" I said.

"Indiana," he said.

"South Dakota," I said.

"Ohio," Kryzewski said. Like just about every coach I'd interviewed
down here, I could see how his competitiveness played out in every
arena, even in the simple act of choosing the right state. "Let's say at
Ohio State that we did what we have done at Duke. There would be statues. There would never be anybody looking for anything wrong with
what you did. It would be a different mind-set from the one we have
within this community."

He was having work done on his house at the moment. Most of the workers were North Carolina and North Carolina State fans, he said. "People don't understand the magnitude of the isolation here. We have to do without that support. And actually, there are eyes close on you, looking at you, not wanting you to do real well. It's toughened us up. We're never going to lose because we have too many people patting us on the back. Never."

He told me a story of how years ago, his oldest daughter, Debbie, had come home crying from Northern High School one day. Some of her fellow students had been following her, taunting and teasing her because of her dad. "It astonished me," Krzyzewski said, "but it didn't completely surprise me. It made me think: But that happened to my mom and dad. And my first reaction to her was: Show your toughness. Maybe I should have showed more compassion. But it was a way of saying: You know what? We're Polish, we're from Duke. And you can't do anything to hurt us. At the end of the day, we're who we are, and we're proud of it. And whatever you do, it's not going to stop us from being that. That's one of the reasons my daughter Debbie and I are so close. Because she went through more of that crap than the others. We hadn't established ourselves. My other two daughters were able to respond from a position of strength: Look, you can say what you want, but we're a national championship program—that type of thing."

In a sense, it was Krzyzewski and family against the world. There was a 1940s feeling to the coach—the Polish squad leader of a bunch of American types: country boys, city slickers, smart guys, dumbasses. He believed in tough love and duty. It was your platoon dug into your foxholes against the oncoming hordes. The world was divided into us against them. His wife, Mickie, was a Southern Baptist from northern Virginia. They married on the day he graduated from West Point in 1969. She shared with her husband the sense of family as us against the world.

A story recounted by the sportswriter Bill Brill is instructive. At the ACC tournament finals in 1991, Duke was playing North Carolina. Featuring Christian Laettner, Bobby Hurley, and Grant Hill, the same Blue Devil team would go on to win the school's first national championship, but they were having a hard time of it that afternoon.

At the time, Krzyzewski's daughter Debbie was a sophomore at Duke. She was going out with Brad Evans, a sophomore at North Carolina. Before the game, Mickie had laid out the ground rules to Brad for sitting with the family. "You are sitting courtside in the front row of the Duke section with the Krzyzewski family," she explained. "There can't be any outward displays of support for Carolina, not as long as you're with us. This is family."

Family is family, but this was Duke versus North Carolina. The Tar Heels were in the process of scorching Duke; the final score would be 96 to 74.

The Tar Heels' Rick Fox punctuated the rout with a rim-rattling dunk, and the unfortunate Brad could not help himself. He clapped. A single clap, apparently, but the thwack could not have been more resounding had he fired a bazooka. Debbie jabbed her boyfriend in the side and the two began to squabble, shortly thereafter to be separated by Mickie's sister, Donna, who proceeded to sit between Brad and Debbie.

Mickie steamed. She decided that the family would leave the game before it ended. "I'm not going to give Brad the satisfaction of seeing his team win," she told her sister.

Mike Krzyzewski happened to glance over at his family during the feud, as he sometimes did during games. And 14 years later, he was not about to let Brad Evans off the hook. "My wife was probably upset that I might have seen that," he said. "He should have the right to . . . but not when he's with us. Not that I'm looking at them all the time, but when I do look there, my wife wants me to see a unified team."

Away games for Duke were now so acrimonious in the stands, he said, that Mickie rarely accompanied the team on the road anymore. "In the Catholic church we have an expression," Krzyzewski said. "Don't put yourself in the occasion to sin." He translated the expression into a dialect spoken more commonly in Chicago: "You know, don't be an idiot."

Family as enclave. It had started with his mother, Emily. By the time she died in 1996, she had seen her son win two national championships, and even now she was never far from his thoughts. Her maxims kept drifting up from his memories into the locker room, where his teams became accustomed to his mom's homespun wisdom.

"Get on the right bus, she used to tell me," Krzyzewski said. "I'd say, 'I know which bus to get on, Mom. I know how to get around Chicago.'"

"She'd say, 'No, no, no, Mike. I'm not talking about the CTA—Chicago Transit Authority. I mean if you get on someone else's bus, make sure that person is good. Sometimes, people driving the bus will take you places you've never been before. And other times, you will drive the bus. Make sure you allow only good people on your bus. And you will take people to the good spots.'"

In his own life, he said that he'd ridden a lot of good buses. The West Point bus. The Bobby Knight bus. The U.S. Army bus. The Duke bus. He'd allowed Shane Battier to drive his bus. Christian Laettner, too.

Until his mother died, Krzyzewski called her after every game.

"One day, she asked me: 'Mike, how did all this happen?'" She meant the championships, the lionizing of her son, the good fortune that had been showered on them all.

Just as Krzyzewski started to tell me the story, he stopped. He cleared his throat. He turned his head to the side. His cheeks flushed. His eyes reddened, and then in their corners, tears began to well. "I'm going to start crying," he said.

He cleared his throat again and went on, his voice thick with emotion. "She asked me how it happened, and I said, 'Because of you.'

"She said, 'C'mon, c'mon.'

"And I said, 'Mom, if I could do as well in what I do with the resources I have as you did with what you had, I'd be unbelievable.' I said, 'Mom, you did better than me. I'll never be that good.'"

The planes of Krzyzewski's face were lit by various intensities of red. "She's the reason I'm not afraid to lose," he said. "That's because I always had her. I didn't know I was armed this way until I was well into my adulthood. And you know what she wanted from me? Nothing. I tried to give her money and she'd give it back to the kids at Christmas. She said, 'Mike, I got enough.' C'mon . . . who is like that? She's a saint."

I knew what he meant. I told him about my own mother, her generosity and kindness. I decided to leave out the parts about her feelings towards Duke and Krzyzewski. Why ruin a husky moment?

"Then you know," Krzyzewski said. The whole exchange had a kind of lull-in-battle intimacy. The tiny office could hardly contain the rawness of emotion. Here was the man who had come to represent everything on earth I loathed, and he was crying in front of me. We were woo-wooing about our mothers. The world was too damned complicated. Man, this hatred gig was tough.

Not that I knew enough to like the man; far from it. But I'd glimpsed the vulnerabilities beneath the steel-plated armor. And I realized that in my own family, I had been Duke—hot-tempered, an outlander, descended on my mother's side from recent immigrants—to my father's North Carolina. And for the present, anyway, I was a city dweller. And in the heat of games, I resembled the profane Krzyzewski far more than I did the calm and circumscribed Dean Smith. Philosophy and reflection could wait until after the game was over. Just win, baby! In his views, Krzyzewski was clearly opposed to most of what I believed. But in his tone, he was my shadow brother. This was spooky indeed.

Krzyzewski pulled himself together when I asked him why Duke was the most hated team in America. (*Bye-bye, mamas.*) I proposed that the Eighties were Duke's decade. They were America's team. Back then, the Blue Devils were like boy bands: squeaky-clean, cute, irrepressible, mostly white (with an occasional black guy like the point guard Johnny Dawkins singing lead). They played hard, they played together, and they had the added advantage of being the underdog, the team on the rise. They were the perfect embodiment of the Reagan years: led by a conservative, cranking out the P.R., representing an institution peopled by those unapologetically in pursuit of personal success.

The Duke of that era finally engaged in its epochal confrontation with the forces of darkness (i.e., a predominantly black team) in 1990 and 1991 when they took on the University of Nevada, Las Vegas, in the Final Four. In the first game, they were slaughtered. In the semifinals the following year, they upset the previously unbeaten Running Rebels and beat Kansas (and Roy Williams) in the finals to win it all at last.

In the Nineties, by contrast, Duke became the team America loved to hate. How had this happened? If I may venture a proprietary guess, the answer had to do with the mutterings and complaints of college basketball fans all over the country who discerned a double standard being applied to the Blue Devils. They played physical defense and got called for fewer fouls than their opponents, most notably against Maryland in the 2001 Final Four. There was the Duke sanctimony—the holier-than-thou attitude, the "We're scholars and basketball players, unlike the rest of you hired hands." And they were slavishly praised by lapdogs like Dick Vitale. What made Duke so special? Over time, these solitary judgments had coalesced into a popular verdict: Duke was America's most hated team.

How did Krzyzewski explain this?

"In the Eighties, we hadn't won yet," he said dryly.

So it's just winning? I asked.

"In the history of sports," he said, "if there's a team that wins a lot, a certain percentage of people are going to want them to lose. I understand it."

And you believe that explains the degree of hatred? I asked.

"*Hate* is a strong word," he said.

It seemed an accurate one, I said. I was speaking from experience.

"I guess in a sports way, they hate us," he acknowledged. "We're not a state school. We don't represent the masses." So here was class again. But, oh, the ironies: Krzyzewski had scrapped his way out of Chicago, a son of the working class, hyperaware of every potential slight, and he had ended up doing the bidding of a private college that had come to represent rich kids and wealth, despite that institution having begun as Normal College, an impoverished Methodist school in nineteenth-century North Carolina. The only thing that was normal here was the degree to which the integers of class and privilege in America got all scrambled. Suddenly, I saw Krzyzewski as an Indian fighting for the British Empire against the local settlers.

So how does all this hatred affect you? I asked.

"It makes me want to win even more," he said.

There was yet one more irony to broach. It had been increasingly said—I'd heard the notion entertained by the writers Caulton Tudor and Bill Brill not long ago—that Mike Krzyzewski was turning into Dean Smith, his longtime archrival, not that either coach would give the other the satisfaction of admitting it. Krzyzewski cloistered himself from the press these days. He engaged in psychological warfare. He talked about how the game was larger than anyone in it, himself included. And, yes, he won a lot.

He'd also learned to let go of losses like Smith. "Otherwise, you'd go nuts," he said. "Losses stay with me until the next game. But my faith is that my God doesn't do anything that doesn't have something positive in it. The celebrated wins are what everyone sees, but the tough losses are the biggest obstacles to continued success. But if you let them stay with you, you're living in the past." Independently, Krzyzewski had arrived at Smith's notion of "the power of helplessness."

Still, that he was evolving into Dean Smith seemed especially ironic; according to Brill, Krzyzewski had once responded to one of Dean's gambits by saying, "If I ever act like Dean, shoot me." "Krzyzewski thought Dean was as phony as could be," Brill said.

It had seemed to me that Dean Smith and Mike Krzyzewski were like two magnets. Unable to simultaneously occupy the same space—that of the classy college basketball power that did things the right way and still won—they repulsed each other. So I wasn't sure how the idea that he was becoming the new Dean Smith would strike Krzyzewski.

Perhaps it was the fact that Smith had been retired since 1997, or maybe it was Krzyzewski's sense of himself as an elder statesman of the game, but he did not initially reject the notion. "I think there are a number of things that he did that I'm doing. Not necessarily how we coach offense or defense. But inherently, there are certain core principles in which he and I are very similar. I admire his competitiveness over a thirty-year period. I've been competitive for a long time like that and I know about what goes into that. I love my kids. I love my school. I love my community. He was devoted to all those things.

"As a coach coming up, I got the chance to compete against somebody . . . well, I understood who he was. And I wanted to beat him and his team but I also respected him and his team. We made each other better. And he built something that lasted. And now I understand him even more. You know, if you're going to do this culture of excellence thing right, it takes time and effort."

And Krzyzewski was near 60 now, starting to think about the sort of legacy he would leave. He wanted the Duke program to not only survive his eventual departure, but prosper. This meant looking beyond individual games to the tenor of one's program, what it stood for and how it would bequeath those values to succeeding generations once the original combatants were gone. In a sense, it was my father's dilemma writ on a basketball team. And in these thoughts of continuity and preservation, Krzyzewski saw the ground he and Dean Smith had in common.

And yet, a residue of the old battles—the "Fuck you, Deans"—hung above the field like gun smoke. The year before, with Chris Duhon and Luol Deng, Duke had enjoyed a stronger team than they had this year. And the week of the first Carolina-Duke game of 2004, Smith had let it slip to the press that Duke was so good that he thought they could go undefeated the rest of the year. Veteran Dean watchers sensed a double intent to the observation. Yes, Duke was very good indeed, so why not place the onus of expectation on them? Pressurize the rest of the season?

Krzyzewski smelled a rat, if you will. At his press conference that week, he responded with a knowing laugh, "Oh, I bet Dean does." The reporters laughed.

"How about that remark?" I asked Krzyzewski. "You seemed to regard it as psychological jujitsu."

Krzyzewski smirked. "Well, that's his shtick, you know. Over the years, I always thought it was funny. Because by doing that sort of thing, he was mentioning the other school." In other words, he was paying a compliment to his opponent, enlarging them in the imagination of fans and players.

"I knew exactly what he was doing," Krzyzewski said. "But being from the North, I'd rather just tell you exactly what I think. That doesn't mean Dean was lying. But he played those games. I'm not a big gamesmanship guy. I'd rather say: Look. I'm gonna try to beat you. You try to beat me. And let's see who wins. And maybe I'm not smart enough to do the gamesmanship thing." He grinned. "Maybe that's what it is."

As our interview neared its end, I had to admit: Winning seemed to keep a man younger than his years. Victory toned the skin, thickened the hair (or maybe in Krzyzewski's case, dyed it; we weren't going to talk about that), and lent vitality to the manner. But there was also something anachronistic about older men who had to win, even if winning was their profession. They may have worn fancy suits, but coaches fought each other with the same bare-knuckled ferocity of kids brawling on the playground. They used the available weapons: strategy, tactics, wit. They careered around the sideline, screaming, staring, kicking chairs, cursing, stomping, sulking. It was such common behavior that its inherent strangeness—the eternal-toddler aspect of it—disappeared in plain sight.

Winning was addictive. And yet, there was something foolish about that craving and its undignified postures, at least to the extent that mastering loss by learning to live with it becomes an essential lesson as one ages. Or, then again, maybe you could just win a lot.

Krzyzewski listened to me go on about winning and losing. Then he said, "I believe in going on to the next play. I'm on a continuum. For me, there's never one event that's an ending."

This seemed like an optimistic note on which to depart. We shook hands and I went back to Chapel Hill.

SHE SEES K AND GOES, "UMMMMM . . ."
Duke versus Virginia Tech
Blacksburg, Virginia
February 17, 2005

In the midst of my sojourn among the Dukies, I sat down to watch a game with my girlfriend, who was visiting. Playing against Virginia Tech, Duke's Daniel Ewing made a nifty behind-the-back dribble to free himself for a layup and I involuntarily exclaimed, "Beautiful move!" My girlfriend, next to me on the sofa, observed: "You never would have said that two years ago!" She was gazing intently at the side of my face. I could feel it. She might as well have been accusing me of infidelity.

I stared straight ahead. Duke looked beatable tonight. The score was 54 to 49 in favor of the Blue Devils with 9:55 left in the game. "Who do you want?" my gal asked. Ordinarily she wouldn't even have asked this question. This indicated that she was on to something.

"Virginia Tech. Of course."

I tried to sound convincing. She was clearly worried that I had altered overnight in some revolutionary way that might portend changes all around. Like IRS agents auditing returns, couples scrutinize each other's little moments, able to read vast amounts into an offhand remark, a choice in clothing, a liking for a new food. What could it have meant for me to compliment Daniel Ewing? What had I given away with my unguarded remark?

The camera lit on Krzyzewski stalking the sideline. "*Umm umm umm,*" my girlfriend said, a low moan in her throat. And so it began. If I was going to insinuate uncertainty into her world, she was going to strike back in time-honored fashion. We had debated the issue of K's attractiveness in the past. I could certainly see his resemblance to a rat. Hence his nickname, "The Rat." And even though I like animals and often find them handsome in their animalness, it was hard—downright incomprehensible—to see how a man that so easily summoned up the image of a rat could be considered attractive. To share a sofa with someone whose worldview was so different from my own was troubling.

"You don't really think he's good-looking," I said. "Even you aren't that crazy."

"What do you mean?" she said. "I'm sure I'm not the only girl who feels this way. He's got something going. Maybe it's a little nasty thing. He should stop dying his hair. And that comb-over has to go. But . . ."

"If that guy sat down next to you on a bar stool, you wouldn't give him the time of day," I said.

"Sure I would. He's cute."

"He may be a lot of things," I said, "like the spawn of Satan. And if that's the kind of look you're after, all right. There's no shame in that. But he ain't cute, my dear."

"Okay. Maybe not cute. But he has something. I would sleep with him in the same way I would sleep with William Buckley."

"Oh, God."

"I would."

As for the other game, the Hokies, playing on their home floor, were bouncing the Blue Devils around as if they were a bunch of sissies. Virginia Tech even sent a football player into the game, the tight end Jeff King. He blocked, hip-checked, elbowed, backsided, and dominated the Duke players at the line of scrimmage. Tech's inside game was primitive and piggish, a mud-wallow lacking all elegance, but it had given the Hokies a chance to win. For the final nine minutes of the game, the teams traded the lead back and forth. Then, with 14.6 seconds left, Virginia Tech's Zabian Dowdell answered a JJ Redick three with one of his own to put the Hokies up for good at 67 to 65. On the other end, Daniel Ewing missed two potentially game-winning threes in succession, the latter as time expired. Duke had now lost its second straight game, and four out of its last seven.

My girlfriend eyed me. "You don't seem that happy."

"I'm fine," I said.

"Fine?" she said. "Just fine?"

"Yes, fine. All right? *Fine.*"

The camera showed a bowed Krzyzewski leaving the court. "Even in defeat he's got it," my girlfriend said. "Ummmmm . . ."

If you were to believe my girlfriend, Krzyzewski wasn't ugly at all. Or, he was charismatically ugly in a peculiarly handsome way. Why did I have to hear this? I'll tell you why. Because the balance of uncertainty in a relationship must be preserved at all cost. If one member of a couple becomes suddenly too mysterious, if he or she betrays signs of a deep-seated shift in allegiance or taste, then strong measures must be taken to restore the status quo.

"This Krzyzewski thing could be a problem for us," I told her. "And we have enough problems."

She just smiled.

THE DOUBLE AGENT

Just when I was beginning to think the Dukies were human, even their coach, I ran across on the Internet a new reference to an article on the Duke–North Carolina rivalry I had written for *Sports Illustrated*. In the piece, I had done a little Duke-bashing, recycled my notion of the school as obsessed with individual success at the expense of public values. Not to mention railing against Krzyzewski, Shane Battier, and the Duke cheerleaders. The fresh link connected me to the syllabus for a class being taught at, of all places, the Duke University Law School. Community Economic Development Law 314.01, to be precise.

Professor Andrew Foster had placed the article on the optional reading list for the first day of class, along with a videotape called "A Day's Work, A Day's Pay," and pages 1 through 12 of the course packet. Okay, it was only optional reading, but still . . . I felt as if my message had silently breached the enemy walls. I contacted Professor Foster and asked to meet him. What, I wanted to know, was he doing with my article?

Foster suggested meeting at a market at the bottom of Airport Road in Chapel Hill, where he and his family lived. When I arrived, he was already there, holding out a cup of coffee towards me. "It'll be ironic for you," he said. "Duke buying you coffee." Professor Foster wore khakis and a button-down shirt, and sported the close-cropped hair of a Young Republican. He also appeared astonishingly young for a law professor, even one who, as it turned out, was not tenured. Apparently, during all these years in which I had been gazing at either basketball or my navel, other folks much younger than I was had been making impressive careers for themselves.

We sat down at a booth among the midday drifters—a familiar realm to me—and Foster began to talk. In order to explain how he'd insinuated my story about Duke and North Carolina into his law course curriculum, Foster felt it advisable to tell me the story of *his* life. He grew up in Urbana–Champaign, where his father taught educational psychology at the University of Illinois.

He remembered his father taking him to basketball games at Illinois, and watching the future pro Eddie Johnson play. When the team returned from the NCAA tournament in 1981, Foster and his father went to the airport to get autographs from the players.

Foster attended high school in Hopewell, New Jersey, graduating in 1987. While there, he fell in love with the girl whom he later married, and with whom he now had four kids. He finished his undergraduate degree at Rutgers in 1991, got married in August 1992, and the following week, he and his wife came down to Chapel Hill so that she could get her masters at the School of Public Health.

"When we moved here," he said, "I was working at Wellspring Grocery, cutting fish, which was probably the best job I ever had. There's a kind of artistry to it, filleting salmon. The easiest fish to cut are the steak fish. Tuna, for instance. It was just fun, cutting fish and talking to people all day."

It was easy to imagine Foster commanding the fish counter. He had the sort of understated congeniality and authority that shoppers appreciate in a counterman. The gals would have swooned over his descriptions of fresh haddock.

"So what happened to you?" I asked him. "You've gone from filleting fish at Wellspring to teaching law at Duke. That's quite a fall."

He laughed. "I'd been cutting fish for about eight months when my wife found an ad in the paper for a job with the North Carolina Associations of Community Development Corporation. I kind of put my hat in the ring and ended up getting the job. That's how I found out about community economic development, the area of law that I work in now. It seemed to match a lot of interests I have in social and economic justice but with a market-based approach to alleviating poverty."

That same school year, 1992 to 1993, coincided with a national championship season for the Tar Heels. Foster's childhood loyalty to the Illini had worn away by then. "We were living in apartments on Rosemary Street right in the middle of town," he said. "Just across from us was a huge Duke fan whom we were friendly with. The night in April that North Carolina won the national championship, the three of us went up to Franklin Street in the rain with blue paint running in the gutters." He chuckled. "It was easy to identify with a winning program. Duke and Carolina are ubiquitous. You're softened up before you get here. You know the teams. You know the players. You know the colors."

His hatred for Duke basketball preceded his affection for Carolina. "I was staying with my wife in California while she was finishing up at Berkeley," he said. "And I was watching a lot of basketball. I never really minded Duke until Christian Laettner stepped on that Kentucky player Aminu Timberlake. Really, other than Grant Hill, I have hated Duke basketball since then. Moving here, it all got exacerbated." This made sense to me. Even his own teammates hated Laettner. No Duke player had ever attracted such animosity.

For eight years starting in 1992, Foster labored for a variety of nonprofit groups, helping poor people by providing legal services and by trying to get banks to give them more loans. In 1997, he entered law school at North Carolina on a scholarship. While he was there, he started a clinic where students could get independent study credit for working with community-development organizations.

After graduating, he began working for Womble, Carlyle, Sandridge and Rice, an area firm, and then struck out on his own in a practice centered around helping nonprofits that were trying to help poor people by assisting their fledgling businesses. Around this time, he received a call from Carol Spruill, an associate dean at the Duke law school. She wanted advice about setting up a clinical course in representing nonprofits. Foster told her he had already drafted just such a plan for a clinic at Carolina. Not long afterwards, Duke hired him to establish the class there. He went home inflated with the news.

"Riley, my oldest daughter, was six at the time," Foster said. "And when she heard I was going to work for Duke, she burst into hysterical tears. I think she still thinks less of me. It was one of the worst things she could possibly imagine."

His wife, too, he confessed, hated Duke basketball with a passion. "I'm not sure she could explain it," he said. "She just hates them. And it's interesting; when at the end of each semester I have all my clinic students over for a party, she can tell right off who went to Duke undergrad."

"A sixth sense," I said. "Some people have it."

"They're different," he said. "Almost without fail, they'll talk about money. They'd want to make sure you knew they went to Duke undergrad. They'll talk about the firm they're going to next. Students who went to other colleges as undergraduates will come in, they'll ask my wife about herself, they'll ask about the kids. This last semester, I didn't have any double Dukes in my class, and my wife said it was by far and away the best group of students I'd ever had."

His study at UNC law school and his experience teaching three courses at Duke law school—the clinic, the classes in community development law, and legal ethics—suggested that Duke students were superior as scholars to those at the UNC law school. But they were also younger, without the breadth of experience and perspective that his former Carolina classmates enjoyed. They hewed to a narrow goal of landing an associate's position with a top big-city firm in New York or Chicago. They tended to lack public-spiritedness. Foster's class in legal ethics, which was a required topic for all law students, allowed him an alarming peek into the conscience of the average Duke law student.

"This semester, we spent a lot of time on Enron," Foster said. "On the role that lawyers played in either facilitating or not facilitating what went on there and in Enron's response to the Securities and Exchange Commission. The Duke students tended to be sympathetic to the role of the lawyer in helping the client do whatever the client wanted to do. They felt it was not the responsibility of the lawyer to impose moral judgment."

"So they feel the lawyer's job is to grease the wheel, not lie down in front of it?"

"That's right. And I disagree with that. They think, *I've got to make money.* They can leave Duke as much as $130,000 or so in debt. They think, *I've got to go up in the firm.* If that means helping clients do what they want even if they don't agree with it, they can live with that. Or so they think. But if that's what you do, you eventually become so unhappy, you can't sustain yourself in your career."

"Because they compromise their best instincts working for their client?" I asked.

"Yes. Surveys tell us that Duke law students leave their jobs within the first three years. And that may be one reason why."

It was becoming apparent that the spruced-up, short-haired, khaki-clad Foster harbored a sly, subversive streak and in his sweet-tempered fashion, stood intent on getting his ambitious students to question their designs for life and career. He presented his convictions as lightly as a waiter laying napkins on a table. And in a small way, that was where my article for *Sports Illustrated* came in.

"On the first day of class," he said, "I try to provide some context for the students, and let them get to know me. It's important that they understand the biases you bring as a teacher. In the past, all I've ever given people is my resume. But Carolina basketball, I love that, and there's the sort of ethos of Carolina generally, its more egalitarian sense of place, and that comes out in your article. The Carolina undergrads love it, and I get to tweak the double-Dukes, which is fun."

The journalist in me was beginning to swell with pride over my small part in the movement for social justice. Screw this basketball, let's reform society, starting with the lawyers! We don't have to shoot them, we just have to get them to care about things! Then, in his unassuming, placid way, Foster took me down a notch.

"Don't get me wrong, however. Mainly the article is just fun for them. I want them to understand that there is more to life than law school and being a lawyer. This is especially important at a place like Duke, which is so competitive. This year's class is absolutely cutthroat. They're the kind of people you hear about hiding books in the library."

Oh. Okay. Forget the revolution. Let's see what games are on TV tonight.

With that, it was time for Professor Foster to head off to a faculty meeting. He left me in an unsettled state of mind. It was always hard to be wrong but sometimes it was just as bad to be right, to realize that the worldview one had grown up with might not be misconstructed. I had often gone off on diatribes against the Duke ethos that were as ill-informed as they were passionate. And man, had those tirades felt good! They cleansed me of vitriol that had been long stored in the rusting barrels of my psyche. Snobbery, elitism, avarice, lack of attention to the local community, an aversion to public values—these were just some of the charges I had hurled towards Durham like mortar shells, mostly in the company of like-minded cannoneers. My recent immersion in the history of both schools suggested, however, that neither Duke nor Carolina had a choke hold on virtue, that both schools had at varying times been forces for enlightenment, both locally and nationally. And at moments, both institutions had succumbed rather too easily to the ugliness of the age.

And yet here was young, fresh-faced Professor Foster, an outsider who'd grown up elsewhere and had sampled the sensibilities of both universities, and he was more or less confirming the understanding with which I had been raised. His views were richly nuanced and profoundly sympathetic. In a country relentlessly committed to the selling of self— every bozo had a blog, it seemed, disseminating *selfness* to a tedious degree—Foster came across as self-effacing. He didn't lob grenades so much as raise questions. His passion for social justice may have tipped the scales a little against the rich kids who were more in evidence at Duke. Still, he saw what he saw.

But I was coming to realize that the case of Duke versus North Carolina wasn't a matter of Democrats (North Carolina) versus Republicans (Duke), nor of liberals against conservatives—as if there were any inherent virtue in either of those positions. Far from it. The argument between the two institutions—and again, it was largely enacted in the symbolic realm where positions detached from their actual origins in history— was between the values of the public sphere versus the private, between American communalism versus American ambition. Of course, most Americans probably believed in a melding of both visions, but in their heart of hearts, they probably tended to favor one or the other. It was a very old argument indeed. How odd that it spilled over into basketball. But it did. Sometimes a game was just a game. And sometimes it wasn't.

I felt the satisfactions of virtue filling me. My gorge began to rise for the first time in close to a month. There *was* a difference.

TEN

The Beast Is Back

I LIKE THE FACT THAT YOU'RE WILLING TO BE MY BITCH

LATE IN THE AFTERNOON ON THE SATURDAY before the second Duke-Carolina game, Melvin Scott and Jawad Williams got baptized at King's Park. They'd been thinking about doing it for a while. So on that day in a room downstairs at the church, in front of their families and friends who were in town for the game, they testified about their new hearts, and then they took the plunge. "Part of me dies today so that a new spirit can be born," Melvin told the assembled. Williams said that "a new beginning is life." Then they immersed themselves from head to toe in the baptismal tank, Melvin wearing socks, shorts, and an undershirt ("don't call it a wife-beater," one of his friends said, "not in church"), Jawad decked out in Carolina gear.

Witnessing the baptism along with Bridget and the other Baltimoreans was Jide Sodipo, Melvin's former assistant coach in high school. At first Jide had been skeptical. "I knew Melvin was going to church," he said. "But going to church doesn't mean you've got religion." But as he watched Melvin and Jawad being baptized, he found himself moved.

"Melvin, you guys did it at the right time," he told them. "Tomorrow the light of God will shine on you. Because you made a stand."

"It might not be the way they play in the game," Jide said later. "Somebody in the crowd might see them in a light they never saw them before. Or maybe they will have a good game, score forty or fifty points, and their future will never be the same. Even though that's irrelevant. They've grown up and know the value of God."

That Saturday night, I was waiting for Melvin at the Smith Center. Enacting the now-mandatory camping-out-before-a big-game experience, a chipper mob of students had somehow pitched tents on the cement terrace outside of Entrance A. Their gathering sounded more like a cocktail party than a beer blast, polite chatter and occasional whoops of laughter wafting across the cool night air to where I sat on an iron railing, waiting.

A guy driving a monster pickup screeched into the parking lot, rolled down the window, and handed a cell phone to a waiting student. "I like the fact that you're willing to be my bitch," the student said.

For some reason, I was happy about that, too.

Just then, Roy Williams arrived in a silver SUV, out of which spilled several young helpers and his wife, Wanda. They were delivering food to the students camped out on the terrace. "We've got chips and salsa here," Williams proclaimed like a waiter at a Mexican restaurant. No one appeared to hear him. Silver-haired and compactly built, he seemed too small to be a larger-than-life coach. The television cameras magnified his ferocity during games and enlarged him into an icon. But he wasn't coaching right now. He wasn't kneeling by the court, imploring his players to get back on defense. Tonight, Little Roy was trudging up the steps of the Dean Dome, lugging bags stuffed with Styrofoam containers of salsa and chips. The students finally spotted him and started to chant: "Roy, Roy, Roy." He explained about the different sauces that were available. Pictures were taken.

While Roy Williams was handing out chips and salsa, Marvin Williams happened to saunter by, basketball in hand, escorting a group of friends and family from his hometown of Bremerton, Washington.

The students spotted him also and switched their chant: "Marvin, Marvin, Marvin." He raised his hand in acknowledgment.

Melvin Scott arrived shortly thereafter to no acclaim at all because he drove his Crown Victoria down the ramp to the loading dock behind the Smith Center, shielded from the fans above. Riding with him was a full complement of Baltimore homeboys, including Jide; his brother, Charles; and his buddy Zeke.

Melvin took the court and began shooting jumpers as part of his standard preparation the night before games. "You don't elevate as high as you used to," Jide said, observing his former charge. "Since your shoulders got so big lifting weights, you don't use your legs the way you once did."

Melvin listened with a neutral expression on his face.

Then Jide decided to play a little defense on Melvin. He casually shadowed him, slapping at the ball, skulking after him from one shooting station to another. "And one!" Melvin finally shouted after Jide fouled him hard on a jumper that dropped through cleanly. This was one of the few hints I'd had that the constant unsolicited advice Melvin received from three-quarters of the population of Baltimore might actually rankle him. For all of the contact, though, he shot better with Jide guarding him, and was dropping more jumpers this evening than the night before the first Duke game.

Jawad Williams, Melvin's roommate and fellow Baptist, showed up, and when he took the basket at the opposite end of the Smith Center, Melvin went down to join him. Jawad shot, and Melvin rebounded. Then Melvin shot, Jawad rebounded. Two fresh-baptized boys, sheened with sweat. They didn't just want to hit the shot, they wanted the shots to go in clean, clean as a purified soul. They frowned whenever the ball drew iron, even if it dropped through the net.

At one point, Melvin drove to the basket and dunked. "The day you dunk in a Carolina game will be history," Jide yelled.

"They got weighted threads in the uniforms," Melvin explained to him. "Weighs you down."

"If y'all lose tomorrow . . ." Jide said, letting the thought hang in the air.

"If 'if' was a fifth, we'd all be drunk," Melvin said.

. . .

At the other end of the court, as usual unable to resist any basketball within a mile radius, Charles Scott was slinging long jumpers from outside the three-point line. Every time he shot, he proclaimed, "Melvin Scott for three!" This was brotherly love. Most of the shots were clanging off the side of the rim, but it was the sentiment that counted.

Zeke joined Charles, and they engaged in a three-point-shooting contest. "Got you, Mr. Charlie," Zeke said, downing a couple from deep.

"Let's run it back," said Charles.

At midcourt, seated on the scorer's table with the Baltimore guys, Jide was now reminiscing about the Duke-Carolina game he had attended in 2001, his first ever. He sat directly behind the Duke bench and caught an earful of Krzyzewski's inspirational rhetoric. "Jason Williams was yelling at Dahntay Jones when they were coming back to the bench for a time-out," Jide said. "And Coach K told them to shut the fuck up. He said, 'You two motherfuckers. I run this team.' Then he told them that Carolina was ready to go down. They were going to trap that poor Adam Boone. Coach K said, 'They suck. They're scared.'" Jide grinned as he told the story. "I was amazed. He was Coach K, this famous coach, and I didn't know he talked like that."

Done with their drills, Melvin and Jawad joined the group. One of the Baltimore guys addressed Jawad, saying: "The Maryland fans are mad at you. They don't like you."

Stoic as usual, Jawad just shrugged, the slightest hint of a smile playing over his face. On the *Inside Carolina* message boards the last two days, the female posters had been going gaga over Jawad's good looks, his smooth-shaven head, his warm eyes, his dark suits, dark shirts, red tie. They'd analyzed his appearance with the same attention to detail that male posters gave to the vicissitudes of his jump shot. "He's more beautiful than Michael Jordan," wrote one gal, applying the ultimate standard in male basketball beauty, at least for a Carolina fan.

The reason the Maryland fans were mad at handsome Jawad Williams was because he had jilted the Terrapins. He had been consider a near shoo-in to commit to Maryland in the summer of 2000, but then Matt Doherty was hired by North Carolina and began wooing Williams with the ardor of a bar-stool Casanova at closing time.

"They say you told them you had a flat tire," Jide said. The rumor circulating through the college basketball community was that Jawad had informed head coach Gary Williams he couldn't visit Maryland on the scheduled weekend because he couldn't get to the airport for his flight because the family car had a flat tire.

"We didn't have a flat tire," Jawad said with a knowing smile. "Dave Dickerson [the Maryland assistant coach] said to me, 'We're sending a limo to your house.' I told him, 'No need to do that.'" Williams had just visited UNC. The battle for his services had ended there and then. "I loved the tradition," he said.

"That's it," Melvin chimed in. "The tradition."

"I didn't have any in-home visits, either," Jawad said, attributing that to living in a tough part of Cleveland. "The coaches didn't want to come to my house."

He stretched out on the floor. Nearby, Melvin leaned against the scorer's table. They'd been baptized, they'd shot their standard allotment of late-night jumpers, and they'd endured four crazy years and now here they were on the verge of their last home game. Days had passed slowly, the years had sped by. Tonight, they were royally at ease.

Melvin offered up his own recruiting tales of dissed coaches. "I fell asleep on Paul Hewitt, dawg. *In the house. On his in-home.*" Hewitt was then coaching at Georgia Tech. Melvin also dodged Jim Boeheim, from Syracuse, who was reduced to walking through the locker room at Southern High School, mournfully asking, "Where is Melvin Scott? Has anyone seen Melvin Scott?"

By now, it was getting on towards midnight. Time to go home. Melvin and Jawad had a game tomorrow. "Security gonna throw you out," Melvin yelled at Charles and Zeke, who were still loudly hooping. Charles refused to quit until he scored again on Zeke.

"Can't let you do that, Mr. Charlie," Zeke said.

While ambling through the tunnel to the car, I asked Melvin how he thought his baptism would affect his play against Duke tomorrow. I wondered whether it might imbue him with a warrior spirit, purify him for battle. Of course, this question may have suggested a rather narrow application of the powers of baptism. Melvin seemed to think so.

"Being baptized makes *everything* I do matter more," he said. Fair enough: He had his religion; I had mine. Tomorrow, at least, it could be boiled down to a three-word creed: "Just win, baby."

Back at home, my mother made her bed on the sofa, laying a sheet down over the cushions. She repeated this process every night, as if her bed was provisional and she a Bedouin in the desert making camp. Sheba, the dog my brother left with my parents 13 years ago for a brief visit, settled on the floor beside her, falling instantly asleep, as she did a dozen times a day.

My mother settled herself beneath a blanket, channel changer in hand, ready to flick her way through the night, as if she could direct her own dreams with the remote control, summoning them from the same studio where Emeril Lagasse kicked it up a notch to the eternal delight of his audience. In the nearly five years since my father had died, she hadn't slept anywhere else.

He'd keeled over from a heart attack in this very room on December 13, 2000, a couple of hours or so after having watched Al Gore gracefully concede the 2000 election to George Bush. "I think I'm having one of those . . ." my father said that December, slumping back on the couch, sliding down into nothingness, breath gone. My mother performed mouth-to-mouth resuscitation. The ambulance took a long time to arrive, or so she thought. He lived in a coma for eight more days. Neurology is still a primitive science in many ways; the brain keeps its secrets well into the twenty-first century. The doctors couldn't tell us whether my father had any consciousness in him during those eight days, but they said it couldn't hurt to behave as if he did. Talk to him, they said. Play him music. Hearing often remains even when a patient is in a locked-in state. A former student of his came and strummed a banjo by the bed. The room swelled with mountain chords and old-time fingerpicking, an Appalachian interlude that would have been entirely to his taste.

I brought in a portable CD player from my car. From home, we retrieved recordings by two of his favorites, Mozart and Doc Watson. Maybe they would spirit him through his long nights incommunicado. Maybe they would bring him home. I remembered my father in the orange armchair at home, directing the *Jupiter* Symphony as he sat there by himself, holding an imaginary baton in one hand, a drink in the other. The drink was real. And so was his jubilation. The house roared with Mozart.

In regard to the CD player, however, I had forgotten that I had left a CD in there, Jefferson Airplane's *Volunteers*. This was a psychedelic record from the late Sixties when (the story goes) the world was going to be remade by free love and drugs and rock 'n' roll. My father had considerable appreciation for love at any cost but not so much for drugs and not at all for rock 'n' roll. The psychedelia of San Francisco in particular would have struck him as a long fall from the heights of Mozart and Doc Watson. In fact, he hated rock 'n' roll and blues and soul and just about any form of music that crackled and popped with electricity. Why, he demanded of us, had we fallen into such savagery when Mozart and Doc Watson were available in the world? (Although it must be acknowledged that he also had a weak spot for belly-dancing music.)

The night nurse on duty punched the play button at about two o'clock one morning as he lay in his coma. It was late December. The night was already long. The cardiac critical care ward was empty of visitors. And out of that plastic box arose the strains of a strident and utopian Grace Slick singing, "Look what's happening out in the street, got the revolution, got the revolution." Overpowering the beeps and pulses of respirators and heart monitors, swirling through the unit, entering the ears of immobile patients, ladies and gentlemen, the Jefferson Airplane.

The night nurse, a country girl from western North Carolina, looked at the CD player and looked at my father—bald, with silvery wings of hair flying out at the side, distinguished even in the stillness of coma—and as she later told us, she thought to herself: He doesn't look like the type. But this was Chapel Hill and she never really knew. The drums and the guitar beat on; Grace Slick kept singing about the revolution, trying to stir the hearts of men and women. She wouldn't have had much luck with my father, that night or any other.

And somehow that country nurse knew it. She listened to the Grace Slick and looked at my father once again and she said, No, it's just not possible. He didn't look like that kind of man. His tastes were written on his face. The country nurse could tell all of this just by looking. And she took Grace Slick out of the CD player and put Mozart in her place. And the night music began all over again, strings and horns, the sweetness of the eighteenth century, grace notes of a different order.

My father would have thanked the nurse, he would have found out where she was from, he would have known someone there.

As for me, the unwitting agent of amplified electricity—at me he would have yelled, "Turn that mess off!" And shaken his head in incomprehension at the uncongenial tastes of his offspring. But not this night. He couldn't speak and we were too far gone in sleep on the floor of the waiting room next to the critical care unit. My father's condition worsened soon afterwards. And then he died. If Al Gore's concession didn't break his fragile heart, having to listen to the Jefferson Airplane probably did. At least, that's the kind of thing you say when someone dies.

At four in the morning, I woke up to a house vaguely astir with voices. From upstairs drifted a brightly lit babble of recipes and sawing and hammering and applause as if the house were full of industrious strangers. Either my mother had fallen asleep with the TV on, or she was still awake, watching. The voices were too frenetic for the hour, and the television's jittery light flashed though the house like an endless summer thunderstorm.

After a time, I fell back asleep.

SEVERAL MILLION MORE LIFETIMES
Duke versus North Carolina
The Dean Dome
March 6, 2005
4:00 P.M.

If Robert Thurman was right about this reincarnation business, I probably added several million lifetimes to my karmic journey in just a couple of hours on Sunday afternoon, March 6, 2005. But not at first. At first, I might actually have whittled a couple of years off my long-term debt. I was down in the tunnels of the Dean Dome in the half hour or so before tip-off and I was all about showing the love, as the young say these days, which in practical terms meant that I was busy shaking hands in an altogether novel way.

As a white man of a certain age, I probably had no business attempting the new black handshakes. I probably should have settled for the straightforward satisfactions of an Anglo-Saxon squeeze. Or, at the very most, a simple high-five. Better to succeed unnoticed at a classic than fail ignominiously at the snazzy baroque of hip-hop. The last time I had the proper sequence of showing love down was 25 or 30 years ago when the expression didn't even exist and I had black friends with whom I slapped hands and traded friendly insults every day in school. My life had become considerably more Caucasian since then. And this meant trouble when it came to fashionable forms of Negro greeting, because, as the late Richard Pryor once put it, "The brothers are always changing the handshake every six months just so white people can't get the hang of it."

But this was the afternoon of the Duke game and everybody was wild with excitement and there was a sense of a battle about to begin and that we were all brothers with the same goal and assembled together in the same place, the tunnel to the court at the Dean Dome. People, mainly the Baltimore crew, all of Melvin's family and friends, kept approaching, looking to give and receive the love, as if we were the oldest of allies. They held their right hands high, then swung them down for a slap that, if executed properly, made a satisfying pop. I loved that pop. But that was only the start of the greeting. After the thwack, both parties shook hands, but not the old-fashioned way, instead with a soul-grip, which was then relaxed and slid out of so that at the conclusion, the handshakers briefly clasped each other's fingertips. Then—and this was my favorite part—each person pulled away with a quick flick to dramatic effect, highlighting the sudden air between them.

Or, they could maintain the clasp and lean in for an embrace. Melvin told me that you did that when you really wanted to show the love.

The choreography was thrilling when I got it right. For one thing, it offered a brief escape from the monotonous rituals of my own class. The

few months of the new handshake had taught me (like a bad dancer) to
let the other guys lead. There was nothing worse than swooping in like
an airplane, hand stretched out wide and high, only to notice that the
other fellow had figured to greet the middle-aged white guy with a traditional handshake. I hated having to pretend to scratch the back of my
head at such moments.

This afternoon, we were all so excited that that we kept giving each
other the full works, a veritable orgy of hand-slapping, finger-pulling,
and chest-pressing love. Charles Scott and I got one down perfectly. "Are
you ready?" he asked.

"I'm ready," I said.

"I think we're gonna get this one," he said. He walked out onto the
edge of the court and settled into his seat behind the bench to watch the
players warming up.

I went and took a seat among my own neighborhood of Carolina fans. "I
have only one suggestion," the man behind me was saying. "And that is
that they do whatever is necessary to stop JJ Redick." He sounded like he
might be proposing something extrajudicial, a sign of the times. I recognized the voice as that of the full-throated fellow who sang out arias of
vehemence every game, mainly to the refs and the opposing team. But
even his beloved Tar Heels felt the lash of his love. One day when he was
excoriating Sean May for soft play, I turned around to see what he looked
like and it took me a while before I realized that he was the pleasant-
looking man who could have been a church deacon—white-haired, red-
cheeked, his body settled in on itself like a comfortable old bungalow.

He and his kind (like me) were partly responsible for my having stum-
bled across an important scientific discovery during the season that I
have named the Law of Inverse Proximity in Regard to Intensity Levels in
the Duke–North Carolina Rivalry. The Law of Inverse Proximity may be
summarized as follows: The lunatic intensity of those involved in the
Duke–North Carolina rivalry increases geometrically in proportion to
one's distance from the court. The closer a participant is to being in the
middle of the action, the less he hates. Though there are occasional

exceptions to this rule; Jerry Stackhouse, for instance, reported to our research team that "I always had it in for Duke. When I went there for an unofficial visit, Coach K told me that I needed to make a decision or he was going to go with Joey Beard. I told him to go with Joey Beard and that I was coming to Carolina. Then I kicked his ass for two years." These sorts of exceptions tended to apply more often to homegrown products like Stackhouse, who grew up inside the rivalry.

In general, though, the players did not feel for each other the disdain that the followers of the teams did. They tended to know and like each other, having played together at camps and in AAU competition during the summers before college. Felton thought Dockery was cool, Redick liked Melvin, Sean May and Shelden Williams were tight and talked every week except the week of a Duke–Carolina contest. The players ran into each other almost weekly at Forty Below, the Durham barbershop where both Duke and Carolina players were teased mercilessly by barbers divided in their loyalties. The players were like members of two neighboring fraternities. Had they not joined one, they would have belonged to the other. They wanted to win the Duke–Carolina game the way they wanted to win every game. They loved the glamour of it, the intensity of this particular battle. But the game was just a game, even if a very big one.

The players' families appraised the games the way a horse trainer studies colts and fillies gamboling in a pasture. How were their boys faring? Were the coaches giving them enough time? Did the offense allow adequate expression of their gifts? Did their sons show enough to make it to the next level? Were they happy? Eating well? The families' loyalty was first to their sons, then the team, then the school.

The coaches wanted to win, of course. But Roy Williams and Mike Krzyzewski were professionals, and they were rightfully wary of attaching too much significance to a single basketball game. This wasn't to say they hadn't clashed a time or two. They each possessed the territorial instincts of yard dogs. During a second-round NCAA tournament game between Duke and Kansas in March 2000, Krzyzewski berated a referee in front of the Duke bench until Williams couldn't take it anymore. It "looked like a coaching clinic with the referees going on over there," Williams later told reporters. Living vicariously through Kansas, North

Carolina fans grew delirious as Williams raced down to the Duke bench, got right up in K's mug, and began yelling. Not one to back down, the Duke coach screamed back at Williams. The two had to be separated and led back to their respective benches. Afterward, Krzyzewski said, "I think he was trying to protect his turf, and I was trying to protect mine." During the 2003 NCAA tournament, before the Kansas–Duke West Regional semifinal game, Williams quipped to reporters, "Duke is a four-letter word, but is it D-U-K-E or D-O-O-K?"

But the two coaches continually reiterated their respect for each other. And neither seemed to get under each other's skin the way Dean Smith and Krzyzewski had. So, again, as predicted by the Law of Inverse Proximity, the rivalry mattered to the two somewhat more than it did to the players but not as much as it did to the fans, whose psychic needs rode on the mercenary backs of the young players out on the court. Broad those backs may have been, particularly in the cases of Sean May and Shelden Williams, but were they broad enough to carry the collective weight of groaning, yearning partisans, particularly the North Carolina fans who had lived too long in exile from their former homeland of Victory? They were tired of waiting for the next game, the next year. As our church deacon said, "I hope they do whatever is necessary to stop JJ Redick."

And so, freighted with such aspirations, the game began.

It was Senior Day at Chapel Hill. Roy Williams continued the Dean Smith tradition of starting the seniors, so little-used C.J. Hooker and Charlie Everett joined Melvin, Jawad, and Jackie Manuel in the opening lineup. Both teams were performing at less than full strength, UNC without Rashad McCants, who was still sidelined with his mysterious disorder, and Duke without Sean Dockery, a key ball handler and defender who had suffered a knee injury.

Duke went up 5 to 0 on an inside shot by Shelden Williams and a three from JJ Redick. Only a minute and a half passed before Roy Williams waved Sean May and Raymond Felton into the contest, replacing Hooker and Everett. Sentimentality was fine; a big early deficit was not. There was too much at stake. With a win, North Carolina, ranked number two in the country, could clinch the regular-season title outright for the first time since 1993 and probably lock down a number one seed in the NCAA tournament. Ranked sixth in the nation, Duke, too, owned aspirations for a top seed. The Blue Devils had also won 15 out of the last 17 games in the rivalry, including the last four in a row.

Within 30 seconds of entering the game, May scored on an offensive rebound. Just as in the February game, possessions in the lane tended to be nasty, brutish, and short. Shelden Williams challenged nearly every shot, with stalwart assistance from Shavlik Randolph. Duke led the nation in shot blocking, and it was easy to see why. A Tar Heel would flash open down low, receive a pass from Felton, go immediately to the basket, only to have the ball frozen delicately in midair by Williams or Randolph. It looked like a reversal of the physical laws of time and motion, a cartoon moment, the ball stuck in the air.

And yet no matter how much the basketball got battered back towards the shooter, Sean May kept emerging time and again out of scrums with the ball securely fastened to his hands. And once he got it, he usually found a way to lay it back on the board. For a player of his bulk, May had a surprisingly delicate touch. Watching him shoot was akin to watching a bear dine on salmon with a knife and fork—such unexpected refinement captured one's notice. His feet were nimble, too. He drop-stepped, pivoted to beat double-teams, and trapped defenders on his backside while he force-marched them under the basket. He was a conscientious student of all aspects of the game, not just its moves but its history.

If there was one player on either side of the Duke–Carolina rivalry who'd come prepared for a maelstrom like this one, it was May. He'd been raised in Bloomington, Indiana, a university town in a state at least as fanatical about basketball as North Carolina. Add to this piece of cultural geography the fact that May's father, Scott, had starred as a six-seven forward on the last team in NCAA history to go undefeated for an entire season, the 1976 Indiana Hoosiers. A businessman with real estate interests in Bloomington, Scott May enjoyed continued access to the Hoosier basketball program and its former head coach, the terrifying Bobby Knight. It would have been easy to presume that son Sean would have followed in his dad's high-top footsteps and enroll at Indiana.

So that when in 2002, Sean May announced his decision to attend North Carolina, the Hoosier state experienced the kind of collective shock that Egypt might were that nation to awaken one morning to find that the Great Pyramid of Cheops had been spirited away in the middle of the night to Monte Carlo.

May, then 17, was nearly run out of Bloomington. Teachers at Bloomington High School North snubbed him; fans screamed insults on the street. Elaborate conspiracy theories about his selection circulated on Internet message boards. May himself emphasized his desire to escape the long paternal shadow and carve out a new identity for himself in Chapel Hill. "I knew that by going to Indiana, I'd be compared to my dad," he said. "I love being his son. There's nothing better. But we're two different types of player. He was more of a wing player, a shooter, and I live more down on the blocks, in the interior. For an eighteen-, nineteen-year-old player to go through the comparisons every night . . . I'd rather go somewhere my father was known but not so well known. For me, going to Bloomington would be like Michael Jordan's son coming here. He would be forever overshadowed by his dad."

When Sean told his father he'd decided to commit to UNC, Scott May told his son that "I probably would have done the same thing."

While the commitment frayed the family's ties to Indiana at least temporarily, the connection to Bobby Knight remained strong. Indeed, Sean May had grown up with plenty of contact with Knight, whom he regarded as "a misunderstood man." "He's a great teacher, a great philosopher of the game," May said. "I've seen him embarrass players, just embarrass them. He tears them down to build them back up, to make them stronger men, and in the old days, guys could take that."

May said that these days, players were "a little more sensitive. They have more integrity for themselves. They don't want to be embarrassed in front of their teammates." He spoke from experience, having felt the sting of Bobby Knight's verbal lash. He was 12 or 13, attending Knight's basketball camp. One day, the coach stood in the paint, giving his daily talk. Seated around him were five hundred campers.

"Where's Sean May?" Knight asked.

May stood up and raised his hand. He was excited, expecting the coach to praise him for his play.

Knight said, "I want everybody to look at this kid, Sean May. He's lazy. He'll never be half the player his father was."

May went home in tears. "I told my dad about it," he said. "I said, 'I can't believe he did that.' My dad said, 'He's really trying to tell you he sees something special in you. But he doesn't want you to know that. He wants you to figure it out for yourself. He wants to know if you're going to be embarrassed and wilt or come back and be a good player.'"

According to Sean, his father "totally agreed with Coach Knight. He said, 'Son, you've got to go out and work harder. You're not working hard enough.' That's when I started every day doing drills for catching the ball, for scoring in the paint, and for jump shots."

In one drill, Sean would line up facing a wall. His father would then call out his name, the signal for Sean to turn and face the pass, while simultaneously blasting the ball at his son's head. "He'd wind up and hit me in the nose," May said. "Eventually he said, 'You're going to learn to catch it or it's going to hit you in the nose every time.' When I learned to catch them, he started throwing high passes, low passes, throwing it *before* he called my name so that the ball was right in my face when I turned around."

As a result, May's hands were famous, as insurable by Lloyd's of London as Betty Grable's legs or Dolly Parton's breasts. Pitch a pass into the lane at an odd angle or an absurd height, fire it at top speed into a tangle of bodies, and somehow May glommed on to it like a shortstop with a glove of the tenderest leather.

The relationship between father and son that begat such hands was old-fashioned; the authority flowed from the top down. But if this fostered resentment, it wasn't visible, perhaps in part because the elder May tended to lurk in the background, letting the spotlight shine on his son.

After every game, Scott May was the first person to whom Sean spoke. "And if it's a home game," Sean May said, "we go out to a restaurant and talk for a couple of hours." There had been plenty for the two to talk about in regard to May's and his teammates' games against Duke. During May's three years at North Carolina, the Tar Heels had managed precisely one victory in the series, and that came in a game in which May had been side-lined by a broken foot. In the first Duke–Carolina matchup of 2004, May had frantically corralled 15 points and 21 rebounds. Unfortunately, many of those boards came on his own missed shots. He was caroming the ball off the backboard at angles of the sort favored by desperate pool players.

After the heartbreaking conclusion to that game—Duhon streaking unmolested down the court to sink the winning layup—May and his father took a corner table at Ham's Restaurant on Franklin Street. "My father told me, 'Sean, you're making it too hard on yourself.' He said that sometimes it looked like I was thinking too much. He also said that it was the first game he'd seen me so relentless on the backboards, just going after everything. But he said that I couldn't just wait to do this only in big games. It's got to be every night." In the next game against Duke, he still shot poorly—14 points on six of 17 from the field—but he hit the boards hard once again and finished with 15 rebounds in another losing cause.

This afternoon, as in the last three games, he was facing Shelden Williams, the six-nine Duke center from Oklahoma, who'd blocked five shots and stolen the ball five times in the game at Cameron. Williams and May were fond adversaries from their days on the high school summer circuit. But their college rivalry was different.

"Those summer games were more one on one," May said. "In college, there's an emphasis on team defense. Shelden and I guard each other very well, but in these games, things are complicated by guys doubling down or coming from behind you to block the shot. It means you've got to look both ways."

May weighed the Duke–Carolina rivalry from several perspectives at once. "We take the Clemson game as seriously as the Duke game," he said in a standard deferral to coaching boilerplate. "But on the other hand, the Duke game is like a season in itself. The stakes are higher. The two regular-season games are must-wins. No matter what happens, the rest of the season feels a lot better if you have those games. There is no better rivalry in sports."

This afternoon, May appeared determined to write himself in capital letters into the annals of the rivalry. His play of late had been particularly ferocious—seven straight double-doubles through the meat of the ACC schedule. In the last game against Duke, he'd scored 23 points and collected 18 rebounds. He'd also garnered an unlikely badge of honor in the internecine warfare between Duke and North Carolina. Of all the players in college basketball that year, May was Mike Krzyzewski's favorite, including members of his own team. "He's the best," he said. "I like the way he plays. I admire his hands. And besides that, he's probably averaged twenty-two rebounds a game against us."

But as well as May was playing in the early going, his team was still struggling. The Blue Devils were negating the Tar Heels' burly inside presence with their usual outside onslaught, having fired their way to an early 19-to-11 lead. JJ Redick showed off a variety of jumpers from all over the court, despite the concentrated defense of Jackie Manuel and David Noel, who were taking turns guarding him. "He hit some tough shots in the first half," Manuel said later. "I was surprised to see some of those shots go in." On one shot, he launched a jumper while falling backwards, impossible to defend. The shot whistled through the net. For the season, Redick was averaging 22 points and almost four threes a game as a leading candidate for ACC Player of the Year.

UNC came roaring back, ratcheting up its defense and harassing Daniel Ewing, who in Dockery's absence had become the Blue Devils' main ball handler. The Tar Heels tied the score at 19 on Jackie Manuel's tomahawk dunk over DeMarcus Nelson, and for a while, the teams traded baskets. But by the break, North Carolina had gained a 47-to-41 margin, its largest lead of the game so far. Redick had already scored 17 points, hitting four of seven from three, and Sean May had played even better, racking up 17 points and nine rebounds, proud statistics for an entire game.

Only one of the two would score again, however.

Throughout the second half, I held my breath every time JJ Redick shot. There was just something so fundamentally right about his jumper that even his misses were satisfying. Actually, his misses this afternoon were especially satisfying. Redick shanked the first shot of the second half, and he just kept missing them. He didn't ordinarily shoot that well against Carolina, a credit to Jackie Manuel's and David Noel's defense, but his cold second half was still a shock. When he missed, North Carolina fans felt spared an execution, as if they'd already been standing blindfolded in front of the wall when the last-minute reprieve came in from the governor. Redick's misses were reprieves. Even a partisan could admit that.

Redick appeared to rein in his game a bit. "I don't know what was going through his mind in the second half," Manuel said, "but he stopped being so aggressive."

And yet Duke was no worse for all of Redick's woes. Though UNC was dominating the paint, the Blue Devils retained the outside. Anytime Redick looked like he had so much as a stray thought, the Tar Heel defenders would converge, and he would sling the ball inside to Shelden Williams, who was on his way to scoring 22 points, or to the six-six swing man Lee Melchionni, so open and lonely he looked like he was trying to hitchhike his way across Death Valley in the middle of a burning summer afternoon. Melchionni's shot reminded me of the former Detroit Piston Bill Laimbeer's tiptoe-through-the-tulips jumper, but he kept sinking them.

With ten minutes left, Duke took a 59-to-58 lead on a Daniel Ewing drive. For some reason, North Carolina had stopped going inside to May and suffered through a seven-minute drought from the field. He found the perfect outlet for his frustration by executing a vicious slam in his good friend Shelden Williams's stoic mug, even hanging on the basket for a second or two afterwards in what was as close to an in-yo-face moment as the suburban May ever came. The Tar Heels tied the score at 64 when Manuel, who was having a stellar game, went up high to snare an alley-oop from Raymond Felton and lightly banked the ball off the glass for two.

Then it was time for another Duke run. When Melchionni hit his fifth three of the half from his lonely corner to put the Blue Devils up 73 to 64 with just over three minutes to go, it looked dark for UNC. Death from Death Valley. The solitary hitchhiker pumped his fist. He had the look of an ugly winner.

The beast despaired. Duke was on a roll, and North Carolina seemed fated once again to suffer defeat. I could hardly speak. The entire Dean Dome went as silent as a morgue.

Damn, the beast said.

Darn, the journalist said, nodding his stunned head in agreement.

It would have been one thing had Duke been waxing North Carolina. When your team is getting blown out, it is possible, even mandatory, to assume a philosophical perspective on the outcome. The mind decides to seek its satisfaction in realms far from the court, tidies up its hierarchy of value so that the winning of games brings up the rear, just behind a good, home-cooked meal and just ahead of cleaning the garage. A good shellacking nudges you closer to your wife and kids, to acts of charity and forbearance. Maybe you'll take up Greek, maybe you'll master the Rubik's Cube. Maybe you'll shave that weekend beard or clip your fingernails or get that tattoo of those supposed Chinese characters lasered off. Many forms of self-improvement beckon.

Close games—contests that seesaw back and forth—ah, they are another matter. Such games become life itself. Detachment is impossible when the longed-for satisfactions of victory, simple victory, appear so close at hand. How the hunger for victory narrows the world to a single point, how the attainment of that single point neutralizes the bitter acids of life's disappointments!

But with around three minutes left to play, victory looked unlikely once again. Better to make friends with loss. I began to relax into that posture of acceptance said to be common among the condemned. If only there were a glimmer. . . . No, better not to go there.

Damn.

Down on the bench at that moment, Roy Williams inhabited a drastically different state of mind, overcome by a degree of conviction worthy of a psychic. He leaned into the huddle. Jawad Williams was sitting on the bench with his head down. The coach flicked him on the knee and said: "Get your head up. Don't do that to me. You do what I tell you to do and we're going to win this stupid game."

Of course, this was Motivational Speak 101. No coach in his right mind would assure his players that by the end of the next three or so minutes, they'd be saddled with yet another devastating loss to their archrival. But no, Williams assured me later, this was different. This time, Roy Williams believed with utter certainty that despite being down nine to Duke, his team was going to win. Williams did not consider himself a gambling man—except on the golf course. But in that moment, with the fans attempting to resign themselves to yet another painful defeat at the hands of the Blue Devils, he felt such ease about the contest's outcome that he would have bet any amount of money.

He'd enjoyed a hunch or two like that before. Back in 1991, his Kansas Jayhawks were down 12 to Arkansas, the notorious 40 Minutes of Hell, in the NCAA regional finals. One of his assistant coaches told Williams, "Coach, they're better than us at the top eight spots." Williams fired back: "Don't worry. We play five at a time."

In the locker room, Williams told his team: "It's our ball to start, and we are going to get a layup. Then we are going to go down and get one stop. We're coming back and then we're going to get another great shot. And then we are going to go down and get one more stop. And then we will come back and get a third great shot in a row, which we do all the time. They will have to call time-out."

And as Williams himself said, may he never hit another golf ball if that wasn't exactly what happened—except that on Kansas's third offensive possession of the second half, the Jayhawks hit a three and cut the deficit to five. The Razorbacks called time-out, and as the teams headed to their respective benches, the Arkansas players began bickering with each other. Roy Williams grabbed two of his own players, pointed at the feuding Razorbacks and said, "See, they're already mad at each other. We're going to win this game." And they did.

As I sat in the silent stands and observed Williams preaching to his players during the time-out, I was reminded of the reserve point guard Wes Miller imitating his coach at Midnight Madness back in October. Wearing a suit and an oversized silver wig that clutched his head like some monstrous animal that had pounced from a tree, Miller stalked up and down the sideline, crouching and swiveling, raising his hands skyward and shaking his fists like a Primitive Free Will Baptist sermonizer whose sole notion of salvation lay in accepting into one's life the gospel of help-side defense.

"I don't know who that guy is," Williams said at the time, laughing.

Oh, but he did. The coach could rant and rave with the best of them. He had a flair for the dramatic. How many clipboards had he broken this season? By my count, there had been at least two; one he had tried to break and it had resisted, so he had to work it over until it finally snapped at the conclusion of the time-out.

As parodies will, Miller's imitation had captured something essential about his coach. For Roy was a preacher, though not a proselytizing Christian. Instead, he was a parson of basketball as true deliverance. He saw in basketball the sweet form that life could take when discipline and standards prevailed, when everyone pitched in towards the same objective, when the game itself was bigger than one player's personal ambitions.

And the reverse was also true: that a squad of selfish gunners (like last year's UNC team) engendered chaos and bad feeling. And chaos Roy Williams resolutely detested. He gravitated towards order and stability. "When I find something I like," he said, "I stick with it: the same golf clubs, the same wife," he once said. He called himself corny. I had been at a press conference once where Williams had announced that he couldn't wait to go home and drink a Coke and eat a hot-fudge sundae with nuts.

Roy Williams was entitled to sweetness and order. The coach had endured enough chaos as the son of Babe Williams. An alcoholic married five times who had abandoned Williams's mother, Lallage, Babe had died less than a year before, in May 2004. At his funeral, the coach had eulogized his father by referring to the Temptations song "Papa Was a Rolling Stone."

He told the mourners: "There were three things in that song that reminded me of my dad: The only thing he left us was alone; where he laid his hat was his home; and the third one—he spent all his time chasing women and drinking."

During the months Babe lay dying of cancer and emphysema, Williams would go to Asheville and sit with his father. His father complimented Williams on his success.

"Well, I learned a lot from you," his son said.

"I didn't teach you one blankety-blank thing," Babe said.

"Yeah, you did, because I just saw what you did. And I did the opposite," his son said.

Both men laughed. They spoke honestly to each other during their visits and came to a hard-won reconciliation. "My dad in some ways didn't do a great job with people as he was living," Williams said. "But as he was dying, he did a great job. He came to grips with me. When he died, he had cleared everything off his chest."

Like Mike Krzyzewski, now huddled with his team at the other end of the floor, Williams had come up the hard way. Maybe even harder. His mother worked in a factory during the day, took in other people's laundry to wash and iron at night. The story had often been told of how, once she discovered that young Roy drank water while his friends drank Cokes, she scrimped to be able to leave him a dime each day so that he, too, could drink a Coke. And how Williams now kept a full stock of Cokes on hand, the better to honor his mother's sacrifice. To this day, he revered his mother, who died in 1992. "She was an angel," he said.

A good student, Williams had the chance to go from T.C. Roberson High School to Georgia Tech on an academic scholarship and study engineering, but he had known since ninth grade that he had wanted to be a coach. Buddy Baldwin, his high school basketball coach, was both a North Carolina graduate and a fan. He pushed Williams towards Chapel Hill, suggesting that he play for the freshman team and learn the game the way Dean Smith believed it should be played.

At Carolina, he made the freshman basketball team but quickly learned the difference between being a big-time talent and a scrappy five-ten guard with seven scholarship offers from small colleges. "When I was here, trying to guard Steve Previs, he was five inches taller and thirty pounds heavier, and I knew . . . but I was more realistic. Kids then didn't have people telling them they were ready for the NBA. The times make kids feel more cocky. Now there are five million high school All-Americans; John Thompson, the former Georgetown coach, calls McDonald's All-Americans 'bad hamburgers.'"

Again, Roy Williams and Mike Krzyzewski had shared a common plight; college marked the end of the road for two feisty but overmatched guards. Their willpower, great though it was, could not overcome their physical limitations.

Williams quit playing basketball the spring of his freshman year and began umpiring softball games for the intramural office. He needed the money. For the rest of his undergraduate career, he worked from four in the afternoon to ten at night, four nights and 24 hours a week.

Because the intramural office was located at Woollen Gymnasium, which was connected to Carmichael Auditorium, then the North Carolina gym, Williams was able to stop by and watch Dean Smith run the varsity practice for 30 to 45 minutes at a stretch. Tar Heel practices were top secret, veiled by black curtains drawn across the entrances and closely policed by student managers and campus cops. Interlopers were bounced quicker than Franklin Street bar fighters. But not Roy Williams. Smith allowed the ambitious young man from the mountains to audit his practices. He took notes, which he had kept to this day.

After finishing at North Carolina, Williams returned to the western part of the state to coach high school basketball, but it wasn't long before Dean Smith lured him back to Chapel Hill, where in 1978, he gained a spot as the lowest-ranking assistant coach. In order to supplement his meager paycheck, he drove videotapes of the Dean Smith Show to television stations around the state, and also developed a thriving business flogging UNC calendars. Williams stayed a decade at UNC, where he enjoyed a prime spot on the bench for Smith's first NCAA championship and helped coach an assortment of legendary North Carolina players. As recounted by David Halberstam, he once took on Michael Jordan in a game of pool.

During Jordan's sophomore year, the Tar Heels had gone to Atlanta for a game against Georgia Tech. In the hotel, the famously competitive Jordan was taking on and defeating all comers. Roy Williams stood watching the festivities. "You're laughing, Coach," Jordan said. "Well, I can handle you, too. Come on, take a stick."

What Jordan didn't know was that Williams was an accomplished pool player. The coach whipped Jordan in three straight games. The player said nothing as the players disbanded to return to their rooms for the night. The next morning, Jordan boarded the team bus, coldly shouldering his way past the coaches. "Hey, Michael, what's the matter—Coach Williams beat you in pool last night?" teased the assistant coach Eddie Fogler.

Jordan swiveled to confront Williams: "You told everyone!"

"Roy didn't tell anyone," Fogler said. "I could tell from looking at you what happened. The defeat is on your face."

When Larry Brown left Kansas in 1988 after having won a national championship with North Carolina high-school star Danny Manning, Dean Smith lobbied for the unsung Williams to take over the Jayhawks job. He dazzled his interviewers with a characteristic mix of intensity, organization, and down-home manners. And subsequently rewarded the faith of his mentor Smith and his new bosses at Kansas by posting a winning percentage of .805 over the next 15 years, the highest among active coaches. He took Kansas to four Final Fours and nine regular-season conference championships.

In 2003, after Matt Doherty had been shown the door, North Carolina came calling for Roy Williams once again. Though UCLA had also offered him the head spot, Williams put off all consideration of his next move while Kansas battled at the Final Four, where they eventually lost the national championship to Syracuse in particularly brutal fashion, 81 to 78. Asked about the North Carolina job by CBS's Bonnie Bernstein in the painful aftermath, Williams had snapped on national TV for all to hear: "I could give a shit about Carolina right now."

But that was then. A few days later, he gave a shit. In April 2003, Roy Williams was heading home, even though he would discover that home had changed while he was gone.

In his absence, Duke had become the gold standard for ACC—perhaps even NCAA—basketball. Upon his return, Williams dove into the rivalry. "Having to beat those folks eight miles away is not easy because they have no holes. It gets very difficult to recruit against them," he said.

Interestingly for a Tar Heel native, Williams had not grown up rooting for North Carolina. "I wasn't really a college basketball fan until I was a senior at T.C. Roberson High School," he said. "Nobody in my family had ever gone to college and so I didn't really follow the sport. Maybe the first ACC game I ever watched was my senior year—and I watched Carolina play. And I said, 'Man, I like the way they look. I like that Carolina blue color.' But I did not grow up loving North Carolina and hating Duke. In fact, at one time, I considered attending Duke."

The first campus he ever visited was Clemson's, to see a football game. But the second he stepped onto was Duke's.

"What were the circumstances?" I asked.

"I don't know whether we want this out," he said with a grin.

"Now that you put it that way, of course we do," I said.

"I was on the square dance team in high school," Williams said. "We performed at the Duke Folk Festival, probably in the fall of 1967. Joan Baez was there. B.B. King. We danced two straight nights."

Williams bought a Duke T-shirt at the festival. "I wore it once," he said, "and then I thought, 'Why am I wearing a Duke T-shirt? By then, I knew I wanted to go to North Carolina. So I never wore it again."

He regarded the Duke–North Carolina rivalry as the biggest in sports. "I think Duke–North Carolina raises everything," he said. In recruiting, he believed in talking positively about his school, not saying anything negative about another. However, he made one exception to that rule. "If you use this," he told me, "be sure you make it clear that this is the only negative thing I'll ever say. Here's a statement I made to a prospect that had narrowed his choice down to us and another school. I said to him, 'If you go to school X and you play school Y, the next morning everybody that played on X and Y are going to know that. But if you come to North Carolina and you play Duke, the next morning every college basketball player in America is going to know who won that game and who played in it.'"

The loss at Cameron in February had initially demolished him. I remembered him in the darkened stairwell, sitting alone in a world of hurt. Williams believed that the Tar Heels had had a chance to steal one, despite having played poorly in the first half. "I told our assistants near the end of the game, 'You know, we have a chance to win this blessed game, even though we've played awful.' Then, boom, eighteen seconds later, you don't feel very good at all. The suddenness and the finality of that loss was terrible."

Losing was always hard for Williams. "The lows are much lower than the highs are high," he told me later. "I don't deal with them nearly as well as I talk to my players about them. They hang with me for such a long time. A long time ago, Coach Smith told me that one of his biggest worries about me as a head coach was that I took losses so hard as an assistant. He said they were much harder to take as the head coach, and he was right."

That night on the staircase, he'd been thinking about his impact on the game. "We had the ball with eighteen seconds and we didn't even get off a shot," he said, "and yet two days before, we had gone over the play and the different options, all of our sets, our last second plays. If they do this, we do that. But we just had a total breakdown of play. A lot people will say, 'Coach, are you ready tonight?' and I always say, 'Oh yeah, I'm ready.' But the guys in short pants are the ones that make the difference."

"Did you feel jinxed?" I asked him. I knew he was a superstitious man, a believer in rituals and omens. Since his return, each of the Duke–Carolina games had come down to the last possession. And North Carolina had lost all three.

"I didn't feel jinxed," he said. "I just felt like we hadn't made the plays at the end. They wanted to take away Rashad coming off the down screen," Williams said. "It's my guess that Mike said, 'Hey, let's look out for this kind of thing.' But it shouldn't have mattered because we had five options off that set. If somebody takes something away, that opens up something else."

In the end, however, Williams wanted his players to be willing to free-lance if necessary. "Every play we have, our players have the freedom to break off that play and be a basketball player. Two days before that Duke game, I said to Raymond, 'Don't forget that your penetration is a major part of this play.'"

Williams grimaced. "And he had Daniel Ewing reaching from behind . . . Raymond had him beaten."

But Felton had already made up his mind what he was going to do, Williams said. "I'm always saying, 'Don't make up your mind before the play. In fact, we beat Connecticut the year before on the same exact play and Raymond told Rashad as they were walking on the floor, 'I'm coming to you.' He told me that and I said, 'Raymond, you shouldn't ever do that. Don't make up your mind. Let the play dictate it. Read the play and then make your decision.' I often say to my players, 'if you were on the playground, what would you have done?' And the guy will say, 'I would have taken a dribble and dunked it.' I say, 'That's what I want you to do.'"

That night, Williams went back to Chapel Hill and watched the game tape. When he spoke to his team the next day about the Duke contest, he said, "As a coach, I should have done differently. And Wes Miller, it was your fault for not pushing Raymond harder in practice and better simulating Duke's spread offense and tempo. It was your fault, Sean, because you should have gotten one more rebound. It was your fault, Rashad, because you had a shot that would have put us up and you missed again. I went on down the line to Coach Robinson and Coach Holladay—it was your fault, I said, because you should have done a better scouting report. It was not Raymond's fault, it was our fault. And if anybody says to you why did Raymond do that, make darn sure you cover up for Raymond. Say it was Coach Williams's fault. Say it was Wes's fault. Say it was everybody's fault."

The Tar Heels had won every game since, demolishing most teams with audacious speed, Raymond Felton leading the way.

In the first Duke game, the Tar Heels had played abysmally and still nearly won, and Williams felt that showed how good his team could be. The second Duke game now offered a dwindling chance to actually be that good.

The teams returned to the floor. The score was 73 to 64 in favor of the Blue Devils. During the final three minutes, my notebooks suggest that I was writing in the dark while under the influence of a jungle hallucinogen. The letters grow bigger and bigger, looping off the edges of the page, a mad calligraphy of comeback. To the extent that they are decipherable, here is what the notes say happened in the final three minutes of action. I trust there is some truth in them.

Jawad Williams tips in a missed Carolina shot. . . . Fresh-Baptized. . . . Score: 73 to 66.

Daniel Ewing kicks ball out of bounds. Possession—UNC! Marvin Williams fouled inside. "ALL BALL!" Billy Packer can be heard screaming from press row. . . . ASS. . . . Williams goes to line, sinks both free throws . . . coolly. . . .kid shooting by himself on the playground, pretending to be in sort of situation Marvin Williams is actually in! Score: 73 to 68. 2:03 to play. WHOA!

DeMarcus Nelson clanks the front end of a one and one . . . SAY GOODBYE TO LAST MONTH'S HERO . . . on other end, May scores contested layup (Shelden Williams appears to block first Carolina attempt) and is fouled. Sinks bonus. Score: 73 to 71. 1:44 to go. WOW!

Fans nuts now! Duke gets ball into front court where Melchionni misses a three but Shelden Williams the rebound . . . DAMN! . . . ball to Redick—OH NO. . . . YES! (Redick open three from corner (how he so open?) but shot spins out.) May the rebound.

Felton goes hard towards basket, pull-up floater, ARGHHH! Ball drops in, spins around, and bubbles out! NO! Duke ball. 36.8 seconds. DESPAIR. NOT AGAIN! TAKE UP STUDY OF DEAD LANGUAGES. . . .

Ewing over the mid-court line . . . Noel beaten, slaps ball from behind . . . loose ball! scramble! . . . what happening? . . . players on floor . . . UNC BALL! TIME-OUT! FELTON!

27.8 seconds. 73-71 Duke up two. Felton drives into gut of Duke D . . . foul . . . two free throws . . . hits first . . . 73-72 . . . NO . . . NO! . . . Marvin Williams! . . . WOW! SHOT! GOOD! AND ONE! OH MY GOD! . . . (Felton? Or someone tapped his missed FT towards Marvin . . . he got shot onto backboard for two!) Now free throw! GOOD! SCORE: 75 to 73 UNC by 2! Time-out. . . . 17 seconds . . . like last month's final possession, only teams reversed.

17 seconds . . . Duke up court . . . Redick off double screen . . . open look from 3! . . . OFF! ball to Ewing . . . shot . . . MAY! REBOUND! GAME! 75 to 73, FINAL! CAROLINA WINS! FELTON FLINGS BALL TOWARDS RAFTERS!

There had to have been something more to life than basketball, but in those moments after Redick missed the three and Ewing the two and Sean May snared his twenty-fourth rebound to go along with 26 points and the game ended, I could not have told you what it was. I had merged into a mass of ecstasy 22,125 souls strong. We screamed until we couldn't scream anymore, and then we started screaming again. I was so light-headed from yelling that I nearly fainted, staggering into the screaming fan in front of me who thought I was trying to hug her. She turned around and hauled me to her substantial bosom and nearly squeezed the life out of me, so happy was she. Which meant that I nearly fainted again, like someone being administered the Heimlich maneuver as a cure for stark, raving joy. Mind you, I tell you this in the service of a thoroughly objective account of nonobjectivity.

I had thought I might be beyond this sort of thing.

The students, many of them painted blue, had stormed the court, forming a writhing blue mass. On the floor underneath them, it was possible to make out a clump or two of Carolina players, one of which seemed to contain Melvin, Jawad, and Jackie Manuel, the three scholarship seniors who'd begun their career here with an 8-and-20 record, the worst in North Carolina history.

"Folks, please, I like the fact that you're out here, but please get off the floor," Roy Williams begged the crowd, microphone in hand. "We're going to have a party. We're gonna cut down the nets. We need you out of the way."

Melvin climbed the ladder first. At the top, he trimmed a sprig of net and then pantomimed a little dance number with his arms alone. He twirled his piece of the net at the crowd and everyone roared. He was followed by his fellow scholarship seniors, Jackie Manuel and Jawad Williams. And then the rest of the team.

Rashad McCants lugged himself up the ladder to get his piece of net. His necktie hung askew over his white dress shirt. He looked like he'd been at a bachelor party the previous night.

The seniors then spoke to the students. Wearing the net on his head, Melvin said: "First, I'd like to thank God. Then I'd like to thank my mom. Otherwise, I wouldn't be here. And thanks to the coaching staff for putting up with me because I'm so crazy." He'd played 22 minutes and missed both of his shots. However, he'd pulled down three rebounds and made a block, an assist, and a steal.

In the press room, a reporter asked Roy Williams if he believed in defining moments. He answered most poetically. "I believe in luck, the stars and the moon. Everything's defined. Sometimes things happen because it's right that they happen. Jackie, Melvin, and Jawad deserved the win, not only for going through eight and twenty, but for going through some other things as well." I suspected he meant the atrocities of the Doherty period. "Before this, we hadn't done it," Williams concluded. "But in basketball, the more you win, the more you win."

In the players' lounge, Melvin was wearing Samos gear and a Cubs cap. A piece of the net was stuck behind his ear like a cigarette. As usual, he was surrounded by writers. His newspaper quote-to-point ratio was probably the most impressive on the team. He complimented the Blue Devils. "They fought to the end," he said.

He was in an expansive mood. The reporters taped and scribbled his every remark. "This is better than winning by forty," he said. "This just shows that if you hang in, there'll be a brighter day. Not just in basketball but in life. Life isn't easy. I hope when people see me and Jawad giving testimony, they understand what we've been through. I was telling Jawad that if he looked into my eyes and he saw me somewhere else, then I apologize. But it hit me, what this was all about. He was missing shots out there. And I was worrying more about the speech after the game than the game itself. All day, I was trying to be regular, to be calm. When I passed the ball to Reyshawn for that three, you would've thought I had twenty points. I wanted to score twenty points. But to beat Duke—not scoring a single point didn't matter. I didn't want to break down out there. The whole college experience has been humbling. Me and Jackie and Jawad were balled up like yarn down there on the floor. This won't settle us. We want more. We deserve more."

When it was time to clear the players' lounge, Melvin and I clapped hands with a loud *thwock*. Then we leaned in shoulder to shoulder, chest to chest, and embraced. It couldn't have gone better.

ELEVEN

We Must Learn How to Explode!

"Having penetrated, in the course of years, quite deeply into two or three religions, I have always retreated on the threshold of 'conversion,' lest I lie to myself. None of them was, in my eyes, free enough to admit that vengeance is a need, the most intense and profound of all, and that each man must satisfy it, if only in words. If we stifle that need, we expose ourselves to serious disturbances. More than one disorder—perhaps all disorders—derive from a vengeance too long postponed. We must learn how to explode! Any disease is healthier than the one provoked by a hoarded rage."

—E.M. CIORAN

TWO YANKEES IN THE MIDST

"The question," Adam Gold was intoning into the microphone with world-historical gravity, "is whether Roy Williams is in the same category with Mike Krzyzewski." He answered himself: "Historically, Roy Williams is *not* in the same category as Mike Krzyzewski. He *might* not be far off three years from now. He *might* win three championships in the next four years. We don't know. So he *might* be judged as an equal to Mike Krzyzewski. But we'll see what he does next year with a very young, albeit *talented* Tar Heel team."

My skin prickled. The journalist wanted to hear more. He had to admit that there was some justice to Gold's pronouncement. The beast, however, was on the verge of a stomp. What partisan wants to hear reason unless it's his own?

I had stopped by the Adam Gold show, a weekday afternoon sports radio program that ran from three to six and emanated from Raleigh on 850 AM, a station also known strangely as The Buzz. (For some reason, radio stations now had names like silly rock bands: The Buzz, The River, KissFM.) Actual games were merely compelling interruptions in the on-going obsession.

Shave-headed and goateed, wearing a short-sleeved yellow shirt and a black choker—everything screamed, "I'm a radio personality! An iconoclast! A visible guy in an invisible medium!"—Gold qualified his stance slightly. "Of course, North Carolina won't be returning any player as good as Shane Battier, Chris Carrawell, or even Nate James. Nor is there anyone in the incoming class of Tar Heel recruits as promising as Jason Williams or Mike Dunleavy." (*That* I found debatable.)

"And North Carolina will be *lucky* if Tyler Hansbrough [the incoming big man from Missouri] develops into a player as productive as Carlos Boozer," Gold added. Dread arose in me. Next year, Duke would be the juggernaut Carolina was this year. That might not hurt now, but it would sure hurt later.

A commercial came on. Joe, the producer, asked Gold, "Who do we have in the big Kazakhstan–Switzerland hockey matchup tonight?" Outside, it was an enlivening spring day. The green and balmy Falls of Neuse Road bustled with traffic. Inside, we huddled in the chilly bunker of a studio, broadcasting to postapocalyptic bands of ACC fans who were scavenging for scraps of gossip, shards of information, reusable bits of opinion. Our pasty-white cave complexions glowed unnaturally in the dim light. And so did Gold's collection of bobble-head dolls, among them the NC State football coach, Chuck "the Chest" Amato, whose head seemed to vibrate with particular agitation.

. . .

I had expected Gold's show to be the kind of crass, bullying, mean-spirited sports talkfest one heard on air in New York, in which it appeared that the callers were all lone gunmen stewing in their Queens one-bedrooms, just waiting for the right moment to take a potshot at Alex Rodriguez. In New York, the callers and the hosts seemed to actually hate sports, to be trapped in their fandom like couples in a vicious marriage. The call-in shows gave these fans the chance to vent their disappointment in clouds of poisonous rhetoric. One bad game, one awful week, one tragic series, and the athlete was toast. Sports radio had given these fans the means to attain a miniature celebrity of their own. Even if that fame was attained by such pithy summations as "A-Rod sucks!"

Thanks for your opinion, Gary from Kew Gardens.

But in fact, despite a reputation for sensationalism and bias, the Adam Gold show didn't seem particularly incendiary. Before the show, Gold had looked downright scholarly, hunched over his desk, poring over newspapers, magazines, and stat sheets. He struck me as a canny radio vagabond—"I've been fired twice, like everybody in radio," he said—who wasn't partisan himself anymore but made his living inciting partisans. And he didn't have to be a raving nutcase on air to do that. As he said, "All I have to do is have an opinion. Half of the people out there will agree, and half will think I'm out of my mind." The devotees of particular schools cared too much about them to see the overall truth. "Fans," Gold said, "just see from the perspective of their school."

A Jew, a Northerner, and a city guy, from Queens, New York, it had taken Gold a while to feel his way into the circumlocutions of North Carolina culture. And as it turned out, his sense of being an outsider here struck a chord with Mike Krzyzewski. A couple of years ago at the ACC's Operation Tip-Off, the preseason bash for the media and coaches, Krzyzewski approached Gold and told him why he liked his show. They'd both grown up in an ethnic environment, he said, which meant growing up with different challenges. Their early encounters with diversity had shaped their identities. "Krzyzewski said to me, 'You seem more tolerant of ideas.' And it's true," Gold said. "I don't always come out on the side of white, Anglo-Saxon America."

Gold had been rewarded with a trip or two up to the secure sixth floor of Krzyzewski's tower for one-on-one interviews. "He even showed me how he could close the shutters from his control panel," Gold said, sounding as impressed as I'd heard him all afternoon.

At four o'clock, Dave Glenn, the editor of the *ACC Area Sports Journal* (a.k.a. *The Poop Sheet*) arrived in the studio to be interviewed by Gold. In shorts, leather sandals, and sunglasses hanging from the collar of a green polo shirt, Glenn wore the casual look of a man who answered mainly to himself.

He told Gold he was glad he had brought up the comparison between the 2000 Duke team and the 2006 Carolina squad. "That is a perfect topic," Glenn said, "to ferret out the liars and the hypocrites. All fans are liars and hypocrites. They would never apply the same standards to their favorite team that they would to someone else's."

I was beginning to feel a little paranoid, as if these guys knew my private thoughts and were broadcasting a response to them on a very personal wavelength. (The receiver was a chip planted deep within my medulla.) Who was this unreasonable fan of whom they were speaking? *You talking to me?*

Glenn reminded me of an archetypal figure I'd enshrined in my personal hall of American history, the freethinker. Admittedly, personal mythologies of the past are usually suspect. They emerge from wishful thinking. One locates in history what is missing from the present in the same way science fiction looks to the future to discover what can't be found in the here and now. I admit it. Don't shoot. Nonetheless, I seemed to recollect a time (vague as a Thomas Hart Benton landscape of yeoman farmers swinging heroic scythes in golden wheat fields) when every American community boasted, or at least tolerated, its own ornery freethinker. The freethinkers of my imagination loved to take the scythes of their reason and slash their way through the underbrush of humbuggery, the briars and tangles of unexamined custom and dubious dogma. How they swung that glorious bright blade, hacking and slashing and opening up new vistas! The shibboleths fell. Light illuminated the fields. In an exultation of truth telling, the freethinkers spoke their minds. It wasn't that they were against religion, even though religion was often what they cleared. It was that their religion was truth.

In the mush-mouthed landscape of North Carolina sports culture, Dave Glenn struck me as a freethinker. His propensity for telling uncomfortable truths had made him something of a pariah to the sports information departments at the local universities. After the Adam Gold show, we spoke about his niche in the local sports ecology.

Glenn had a lot of credibility in my book. Through dogged reporting, he'd broken open the story of the player and parent revolt against North Carolina coach Matt Doherty's harsh hand in the spring of 2003. Nearly all the state's newspapers and television stations at the time—and most especially the national press, the Dick Vitales and Jay Bilases—had dismissed the reports of the rebellion as exaggerated, and had described the stories of Doherty's temper as normal coaching behavior. They'd reflexively defended Doherty without knowing—or digging for—the whole story.

Not Glenn. He'd gone to players, players' families, and sources inside the UNC athletic department and carefully pieced together an account of a full-fledged player revolt against a volatile young coach. Doherty, Glenn maintained, appeared to have rejected Dean Smith's practice of treating his charges as human beings, not cogs in a basketball machine, and as in the other stories he published in his journal, he laid out the case with a passionate precision befitting the lawyer he had trained to be.

He attributed his perspective in part to his being an outsider. "One of the ways I have a different perspective on the ACC or Duke and North Carolina is that I come from a different place," he said.

He had arrived in Chapel Hill to attend the university (and eventually its journalism and law schools) in 1987 after a childhood in Philadelphia and its suburbs. Right away, he started getting into trouble. In 1989, as sports editor for *The Daily Tar Heel*, the campus newspaper, Glenn wrote a box comparing Duke and North Carolina on the eve of one of their regular-season games. In the coaching category, he gave the advantage to Mike Krzyzewski. He thought the Duke coach had begun to match or surpass Dean Smith in recruiting, and he also believed that Krzyzewski was beginning to show himself a more flexible tactician than his archrival.

"You're a brave man," I said.

"If you're serious about being a journalist," he said with a rueful smile, "you've got to tell it like you see it, even if you're writing for *The Daily Tar Heel*. Let the dominoes fall where they may. Even though I'm now a UNC grad, I've never been a rah-rah guy. My logical lawyer's brain has always enjoyed looking at the whole picture."

The reaction to Glenn's rating of Krzyzewski was immediate, intense, nearly Middle Eastern in its complicated subtexts. "One extreme was Duke fans saying that I was contributing to Dean Smith taking psychological advantage of the situation to kill Duke—that I was actually in Smith's pocket."

Then there were Carolina fans who understood that Glenn wasn't kidding and felt he should be disabused of his views. "Drunk guys left messages on my answering machine, saying 'I'm going to fucking kill you.' They'd say something about 'you damn Yankee coming down here. What do you know about Dean Smith? Does 1982 mean anything to you? Do you see any rings on Mike Krzyzewski's fingers?' Hatred is the right word to characterize the way people feel about Duke and North Carolina down here."

Under his direction, the *ACC Area Sports Journal* covered stories that local dailies wouldn't touch, though Glenn usually used those papers' beat writers writing anonymously. The coverage tended to approximate the bolder forays of a metropolitan daily, a stand that hadn't been appreciated at the sports information departments of the area's universities. Nor, in many cases, by the fans of the university negatively portrayed.

"Given the personal treatment I have received, it's as if I was Darth Vader covering the ACC," he said. "I must be attacked, blackballed, denied credentials if they can get away with it. I'm talking vicious stuff."

The Doherty story ran the week after UNC beat Duke in March 2003.

"Many fans thought that Dave Glenn hates Carolina," he said. "And that he printed the Doherty story to take the big glow off the win over Duke. If the story's ready, I don't care if it's Mother's Day and you're a mother, the story's going forward."

After the Doherty piece ran, Glenn got phone calls that he imagined being made from public phones on Franklin Street after the bars closed. "Some were saying: I know where you live. I know what you look like. Some were saying: How could you do this to your alma mater? I remember one guy said you could end up like so and so. I didn't know who this was. So I guessed the spelling and I Googled it. I found this article about somebody who had been killed.

"When I got death threats at nineteen or twenty, I was ready to call the guy back and tell him to meet me in the alley. But when you're thirty-five and you've got a wife and two kids, it's like: Holy God. I drive slower now. I don't want to fight anybody."

The death threats were only the extreme edge of a nearly unanimous response against the story. "The reaction was ninety-nine to one against it," Glenn said. "It was not pleasant at all. In general, the day-to-day negative reaction I get is not even a bug on the windshield of my world. I'm used to it. This was a little different but I knew that I was right. My wife says I sleep better than anybody that she knows of."

He bridled against a culture he perceived as valuing politeness over truth. "Coming from the South," Glenn asked me, "do you notice how often people introduce things they're about to say as: 'I don't mean to be ugly, but . . .'"

I laughed. I didn't mean to be ugly, but yes, I did notice that.

"Many Southerners feel that in being bold or truthful, they have to apologize." He shook his head in befuddlement. When it came to the local mores, he and I had arrived at a similar place despite my having been a Southerner who'd grown up in the area and Glenn having been a Northerner who'd migrated here. I told him that I thought North Carolina subscribed to a culture of niceness, probably derived from its particular forms of Christianity.

"There's many things I like about the South better than I do about the North," Glenn was quick to add. "Southern hospitality is real; it's not a myth. But I'm not going to apologize for telling the truth in this culture. There are people like John Kilgo of *Carolina Blue* who've never had an original thought in their lives and just sit there and nod and have drool pouring out of their mouth, saying that everything that Dean Smith says, I endorse."

Sometimes, Glenn said, he liked to instruct callers blinded by their partisan views who were objecting to a story he was divulging on the radio: "Go to a message board. Get that view out of your system and don't bring it back into my garden, so to speak. I am the weed killer. You are the weed. I'm going to spray your insanity every time I see it pop up. That is my tiny contribution to the world."

The weed killer's intensity was charismatic, his appetite for truth refreshing. "Intelligence would be a real impediment when it comes to partisanship," he said.

"Touché," I admitted.

"The most beautiful people are the ones who can truly see Duke and North Carolina. They're rare birds. But they're out there," Glenn said. "They're out there."

THE MINISTER OF BASKETBALL

I kept wondering, Why did we obsess over basketball so much down here?

Answers: (a) Because our teams won a lot and winning is fun. (Was it really that simple?) (b) It let us North Carolinians cut loose.

As a corollary to (b), there was another theory: that hatred had its uses, much underrated in a Christian realm (though the ancient Greeks knew the value of a vehement passion like hatred), and that the suppression of good, honest hatred bollixed people up, turned them inside out, and made their niceness somehow mean.

And I know you Southerners (at least) know what I'm talking about—e.g., that sweet lady from down the street whose niceness is a switchblade drawn from her purse as she tells you with concern about the troubles of Ted and Molly across the street, the false compassion dripped over her tale like cinnamon over the foam of her latte.

Somehow, this all seemed the fault of the kind of Christianity practiced within our handsome state. I decided I needed to talk to—of all
people—a minister.

On the Monday morning after the Duke game, a day of high, scudding
clouds, I strolled up the sidewalks along Franklin Street to University
Presbyterian Church for a visit with its pastor, Bob Dunham. Given how
kindly Bob had behaved at my father's funeral, feeding me updates from
the Carolina–UCLA game, I didn't imagine he would turn aside my
request for spiritual guidance related to basketball. And he didn't.
Dressed in khakis and a blue sports shirt, Bob closed us in his office,
which spilled over with novels, poetry, and literary magazines, more
reminiscent of an English professor's quarters than a preacher's.

Here I was in the church of my childhood, the church in which I was
confirmed, the church in which I had fought to stay awake on Sundays
as the customary platitudes fell softly over the congregation like a baby
blanket.

I had been drowsy all those years because church was boring. The Protestant theologians of the twentieth century had somehow reduced God
from a voice out of the whirlwind to a gentle breeze whispering through
the parking lot, from an awesome mystery into a civics lesson, from the
power and the glory to the friendly and concerned. That's if He was
around at all. So that attendance at University Presbyterian Church
struck me largely as an exercise in being good, in should and shouldn't.
You rarely encountered joy or terror. You were rarely if ever possessed
with the spirit. Larger spiritual hungers went unaddressed. Now there
are good things to be said for such moderation in the face of divinity
(the wilds of spirit life teem with their own dangers), but I am speaking
of the bad. This was religion as a Rotary Club meeting. This was religion
as ethical culture. This was religion as a dead magnet with no power to
attract, offering comfort and duty and nostalgia in place of the shock
and disorientation of genuine spiritual feeling.

Or so it seemed to my demanding and bewildered heart. Admittedly, I
was an extremist. I wanted burning bushes, voices from that whirlwind,
visions of ladders to heaven, wrestling matches with angels. I wanted to

know God's true name. As a 13-year-old in the grips of religious despair, I even went so far as to ask Jesus if He wouldn't mind appearing on my bedroom wall right next to the picture of Che Guevara. I promised him I wouldn't get the two of them confused. I had mixed feelings about seeing Jesus up there, among the Day-Glo rock posters and the UNC basketball calendars, but I needed to desperately. If numinous encounters with angels and prophets had happened once, why couldn't they happen again? Why were we fated to live in a posthumous time, stripped of miracles and direct communication with God? Why was I a latecomer to the ball, forced to substitute metaphor for event, interpretation for prophecy, question for answer?

"You'll probably be relieved to hear I don't want to discuss my romantic tribulations," I told Bob Dunham. (He'd been kind to me during a bad time.)

He smiled as if to suggest that I might have been right about his relief. "Instead, I thought we could talk about Presbyterianism and God." His face fell. "Well, Presbyterianism and God and basketball."

Now Bob smiled. Basketball. All right, then. He loved the game. He'd been the sports editor of the Davidson College newspaper back in the Sixties. These days, when Bob drove visitors around Chapel Hill, he took them by the Dean Dome, saying, "Now I'm going to show you the town's real spiritual center. College basketball is the common religion that binds us together."

Having drawn up a loose bill of indictment against the state's Christians in general and Presbyterians in particular, I presented it to Bob to see what he thought.

I told him how I had felt the church had stinted on the ecstatic side of religion, and that while this might be a legacy of Presbyterianism's emphasis on education and learning (several of the early presidents of the University of North Carolina had been Presbyterian), it still seemed unfortunate. And that basketball gave us poor repressed souls the chance to shout and cry and disappear into a mass (a mass!) of emotion. We had heads, but we also had hearts.

Again, it was a measure of Bob's fundamental civility that he listened and considered my notions. "Presbyterianism was the denomination of the middle class," he told me. "The phrase that characterized Presbyterians for many years was their penchant for doing things *decently and in order*. That's how we've tried to govern and conduct ourselves.

"And it is true," he said, "that rooting for Carolina or Duke allows people an ecstatic experience that they don't have in their houses of worship. We yell things that we'd never yell in church, at least those of us who are Presbyterians or members of what are called the mainline traditions."

Bob had a mischievous glint in his eye when he said, "In fact, there are people in this very congregation who are reluctant about going to games because they're afraid they'll get too crazy."

"Anyone I know?" I asked.

"I won't mention names," he said, "but there's one person, a prominent physician, who simply did not go to big games because he was afraid of making a fool of himself. Just too much pressure. He knew that he had to do business with other people who might see him there. He didn't want to lose that persona of being a rational businessperson."

"All right," I said. "So there is a reticence that's all about not seeming crazy. But how about that other kind, that old North Carolina modesty—maybe it's Southern, actually—that aw-shucks veneer that actually conceals pride in one's virtue? How about that?"

Bob answered me by referring to one of those twentieth-century theologians whose preoccupations I had found so unsatisfying.

"Reinhold Niebuhr talked about the insidiousness of the sin of pride," he said. "It's a subtle form of sin, not as visible as sins of passion, for instance. Niebuhr said that the ultimate form of pride is spiritual pride that assumes that not only do you know what God's intent is, but that God is on your side. Then there's the pride of power: You sense you're invulnerable. There's the pride of knowledge: You think you know more than other people, and therefore, you're better. Pride of virtue: that you have better morals than another person and are therefore superior."

"That's what I'm talking about," I said. "That North Carolina modesty—underneath it seems to be considerable pride."

That pseudo-modesty, he said, constituted a form of hypocrisy. "Hypocrisy is often rightly thought of as saying one thing and doing another. But there's another kind of hypocrisy in which we claim to be *less* than we are. There's a kind of self-deprecation practiced by a lot of Southerners that is really false."

Bob then made an interesting geographical observation. "We lived in Richmond for five years, a pillar of civility. And then we moved from there to Alabama, which had no pretense of civility. But in regards to racial politics, the bad stuff in Alabama was right out front. And everybody knew who everybody was, and nobody was pretending otherwise."

"Hatred, but not hypocrisy," I said.

He nodded his head in assent and leaned forward in his chair to evenly convey his next point. "And because of that, despite all the problems Alabama has in other areas, the state worked through a lot of its racism. It's probably made more progress than most Southern states. Because it was all there on the surface. It was so overt."

By contrast, he said that in Virginia, particularly in Richmond, "the civil rights movement happened but it never made the papers. They didn't talk about it." And so racism endured and perhaps even prospered under the cover of silent gentility.

So the lesson of the day, unexpectedly Christian: The outright expression of antagonism was good! Or, at least the *outright* part was good. A good, honest hatred beat a dishonest love. This may not have been Bob's precise take on the matter, I was willing to concede, but perhaps it was close.

So maybe fandom and religious experience were exchangeable currency. Perhaps the agony and ecstasy afforded by sports could actually propel you on your way to hard-won spiritual knowledge that politer forms of instruction (Sunday school, for instance) could not reach. This wasn't necessary to enjoy sports, by the way—the enjoyment was justification enough—but surely the passion aroused by winning and losing filled a hole in many North Carolinians' religious life. Otherwise, life was so circumscribed, so parched of extremity, so comfortably suburban in assumption and experience.

I asked Bob what he thought of this intuition.

"To some degree, I agree," he said. "Zealotry *can* actually be a step on the way toward what I would consider a mature faith. And I'm not saying that basketball fans are immature! I think that for people to get to a point where they can claim the faith for themselves—whatever faith is—they almost have to go through a phase of zealotry. And it doesn't have to be religious."

Here was a minister who seemed to be suggesting that the ecstatic, the demoniac, the roiling of the spirit could be a route towards enlightenment, no matter the venue—a tent tabernacle at revival time or a coliseum roaring with rabid fans. Agnostic though I might be, I was beginning to feel redeemed.

"When my kids were coming along," Bob said, "my son in particular was deeply engaged in sports, and I thought that was a good thing. He would get so wrapped up in it. And over time, such immersion forces you towards a kind of perspective. It doesn't mean you're less zealous about your team but you might become a little more philosophical about their losing. *Through* their losing."

Yes, it seemed that losing in games and in life was a better teacher than winning. At least losing taught you how to lose, and since losing seemed pretty inevitable in one realm or another—like, say, *death*—this was instructive.

Bob said, "When you experience a year like Carolina is going to have next season, you can say, 'All right. We had a great year this year. It'll be okay.'"

"Or like the eight-and-twenty year in 2002," I said.

"Well, the eight-and-twenty year is kind of hard to philosophize away," he said. True. Philosophy had its limits.

"But how about actually hating Duke? Do you not see a conflict between that and living a moral life?"

He laughed. "Ah, hatred. What we call negative emotion."

"Yes. I must confess that from time to time, I am suffused with what we call 'negative emotion.' I sort of lose it. And I have to wonder whether spiritual advancement is in the cards for me. Whether a moral life is actually possible. Because if basketball gets me going *this* much . . . not to mention George Bush . . . and that idiot Aaron Brown on CNN . . ."

Bob listened with what appeared to be a subversive twinkle in his eyes. Yes, he was a churchman, to be sure, but he certainly showed no inclination to occupy the moral high ground. In fact, he seemed kind of happy to sit next to me in the swampy lowlands. Also, I think he knew a little something about watching a coach bait a ref.

"I don't know if I ever told you this," he said. "But before I came to Chapel Hill, I was not a Carolina fan."

"You were a Davidson fan. That's okay," I said.

"I had been at Davidson when they were playing really high-level basketball," Bob explained. He had attended the small Presbyterian school just outside of Charlotte back in the late Sixties, when Lefty Driesell was coaching there and the team featured stars like Mike Maloy and Jerry Kroll. "It was a wonderful time to be there. That team was consistently ranked in the Top Ten. Basketball had a lot to do with putting that school on the map. Well, two years in a row, in the Eastern regional finals, we got bumped off by Carolina."

"Oh, I remember," I said. "We were kids. In 1969, we watched the game at my grandparents' house in Huntersville, not far from Davidson. After Charlie Scott hit that shot, my father and uncle drove us around the Davidson campus so my cousins and my brothers and sister and I could scream out the car windows at you Davidson students. We even made signs. It was fantastic!"

Scott had canned a 20-foot jumper with three seconds left, to beat Davidson 87 to 85. "Now that I think about it, that was *really* nice of my father and my uncle."

Bob winced. Even now, the minister appeared pained by my 11-year-old's idea of rubbing it in. That shot. Memories of such daggers were long in this part of the world. "Charlie Scott, that's right," he said after a long pause. "Everybody at Davidson vilified Dean Smith because he recruited Charlie away from Davidson."

This was true. Scott, the first great black basketball player at North Carolina and in the ACC, had originally committed to Lefty Driesell at Davidson. And somehow, Dean Smith had stolen him away, a theft if you wanted to call it that—and Bob did—that still rankled. "For years afterward, I just could not pull for Carolina. Not until 1982, when they got to the Final Four. By then, I thought that Dean Smith deserved to win a title. But it took me twelve years to get to that point. Looking at Dean Smith up to then, I would just start steaming. Now, once I came to Chapel Hill, and I got to know him and got to see the other side of him—his contributions to his church, his community, and his university—I started to see him in a different light. And I also got to know some of his players who knew him well, and that further altered my view."

Bob swallowed deeply. He had something that he had to share with me. It wasn't going to be easy. "Interestingly," he told me, "I have a similar relationship with people in Durham who feel the same way about Mike Krzyzewski. They appreciate the things he's done for his community, his church. And so, in my more reasonable moments, I could claim Krzyzewski as a first-rate basketball coach and also as a decent human being. As I said, I can be a fanatic. I can live in the world of fandom. But I think there are times when I can transcend that."

He sighed, considering certain hard truths. Transcendence can be difficult. "On the other hand, when he's coaching against Carolina, I just look at him and I start steaming. In the same way I used to about Dean Smith. And I have to admit, there are things that Krzyzewski does that drive me crazy—the way he works the officials out there."

"Oh, I hear you," I said.

"He's sly like a fox. And what disturbs me even more as a Carolina fan is that I see it working."

Y'ALL'S HEADS ARE JUST THE SAME

Charlie Scott, the first black player on the North Carolina varsity, arrived in my life in the late Sixties, around the same time the schools in Chapel Hill were integrated and my siblings and I began to be bused from Hillcrest Circle to Glenwood Elementary School out on the Raleigh road. The bus drivers were usually high school students who wanted to make some money. I remember accidentally stinging one black driver named Eric Cotton on the neck with a rubber band launched from the back of the bus. The rubber band traveled miraculously far, far beyond my wildest hopes, somehow managing to miss row after row of spruced white heads to impact crisply on Eric's neck with a satisfying *whap*. The reaction was instantaneous. He pulled the bus over to the side of the road, choked off the motor, and moved menacingly towards the back of the bus, which had suddenly grown impressively silent. "Who did that?" he demanded.

I believed in civil rights, Martin Luther King, Huey Newton, H. Rap Brown, and James Brown, too. Never in my life had I expected to be on the wrong side of the racial divide. But somehow, accidentally popping Eric Cotton in the neck with a miraculous rubber band had put me there. I decided that the only honorable thing for me to do was to remain utterly silent.

"We ain't going nowhere until I find out who hit me in the neck," he said.

Everyone in the front of the bus was turning to stare at the troublemakers in the back. Polite little girls of the neighborhood, the daughters of doctors and professors and ministers, they eyed us as if we were junior members of the Ku Klux Klan. "I'm not like that," I wanted to say. But I remained silent. I could ride this one out, safe in the anonymous white sea.

"We gonna just sit here, then," Eric Cotton said. "Just sit until one of you confesses."

He was holding the seat backs on both sides of the aisle, blocking the main route of escape. He had an impressive Afro that signaled some new level of forthrightness between black and white. How I admired his stern cool, his phrasing. I was going to steal some of it for use at a more convenient time.

I'll never understand exactly why my best friend gave me up. He was not a malicious guy. Maybe he was just playing, maybe he had a test first period and needed to get to school. Maybe my dear friend was simply honest. "Uh, Wiiii-iiiii-lllllll," he sort of sang my name, turning it into a teasingly long three or four syllables. Everyone on the bus looked at me. Eric Cotton looked at me. "It was an accident," I said.

"That wasn't no accident," Eric Cotton said. "You little boys playing around. You hit me in the neck. I oughta make you walk to school."

With that, he turned and shouldered his way down the silent aisle to the driver's seat. The little girls stared at me with disdain. In the back of the bus, I burned with the embarrassment of getting caught. In the months after that, however, Eric Cotton used to see me getting on the bus and say, "Here comes the rubber band man. Better not see any of that today." It made me feel good, good about myself, good about him, good about the races. I thought that deep in our hearts, we might actually indeed overcome one day as long as we kept our rubber bands to ourselves.

Around that same time, my Aunt Sal told me my head was shaped just like Charlie Scott's. "Really, y'all's heads are just the same," she said. This made me immensely proud. If my thin blade of a head was shaped like Charlie Scott's thin blade of a head, then maybe I would one day play basketball like Charlie Scott. His head, my head were links in an associational logic I've never outgrown. If Scott were a superhero—which to the children of Chapel Hill, white and black, he already was—but if he were a *comic book* superhero, his special power would have been the ability to squeeze himself through small cracks, so fluidly thin was he, head included.

Against Duke in the ACC Tournament finals of 1969, Charlie Scott went wild, scoring 40 points, 29 of them in the second half. Tar Heels won by 11. He was a brilliant player, six-six, a college All-American who could shoot and slice to the basket better than anyone.

A few years later, when I was in high school, I was sitting with my friends Reggie Jackson and Cranston Farrington in the bleachers at Fetzer Field before track practice. Given that Reggie was a shot-putter and Cranston a sprinter, they'd be sitting there a lot longer, as the sprinters and field guys were notorious for lounging around. "Hey, Will," Cranston said. "When you were born, did your mother take two bricks and smash your head between them?"

Reggie grinned. "Who does his head look like? I know. It looks just like Charlie Scott's head."

"But he can't play basketball like him," Cranston said.

THE WARLIKE NATURE OF STONEWALL JACKSON

A few hours after talking with Bob Dunham, I ran into D.G. Martin, one of our leading local citizens, at Branch's Bookshop. In his college days, he'd played basketball at Davidson for Lefty Driesell. "You were a good defensive player," Lefty had told him years later, a compliment he related to me proudly. Because Martin had grown up in a Presbyterian household in Mecklenburg County, the son of a president of Davidson, and attended a Presbyterian college, I elaborated my theory of our passion for basketball as a release from the strictures of Presbyterian modesty and repression. As I spun the idea out, and it didn't take long, I was aware that I might be starting to sound like a monomaniacal crackpot. But that had never stopped me before.

"I don't know," Martin said politely, as Presbyterians tend to do. "I've played basketball against some pretty warlike Presbyterians. And then there was Stonewall Jackson." Martin had just returned from a guided tour of Civil War battlefields in Virginia, so Jackson had been much on his mind of late.

Stonewall Jackson. All right, Martin made a good point. Along with Robert E. Lee, Jackson was the greatest of the Southern commanders during the Civil War, a shabby-dressed, rough-riding, brilliant tactician whose Shenandoah Valley Campaign is still taught in military academies. And yes, he was also a stiff-necked Presbyterian, a moralizing teetotaler, unbendingly Calvinist. And yet was battle not a sweet release for this warrior-prophet? An unbottling of martial spirits fermenting inside all that propriety? Was not the clash of swords and the roar of cannons a welcome sound to such a man? Did not the Bible say that the violent will bear it away?

I didn't know for sure, I was just wondering.

THE MAN WHO LOVED DUKE AND NORTH CAROLINA

The day after Dean Smith announced his resignation in October 1997, Mike Troy was driving up Strowd Hill in Chapel Hill when he had an idea for a poem. He swerved right onto Carolina Avenue, pulled the car over to the side of the road, and there, across from the autumnal Bradford pear trees dropping their leaves in Professor Ruel Tyson's front yard, he proceeded to compose the following, which he now recited to me as we devoured a lunch of smothered chicken and corn bread at a soul food restaurant above the Eno River in Hillsborough:

The Day

We thought the day would never come
So when it came we were all struck dumb
Dumbstruck we all were in this town
When Dean said today I'm stepping down
But when we slept and dreamed one night
We woke and said it's exactly right
To place him there on memory's shelf
It's time that Dean should have himself
A thing begun must always end
For then the new thing can begin
In each beginning is dawn's light
As day must always follow night
Life teaches us all things are one
It's clear when all is said and done
Dean wasn't teaching basketball
He taught us this: "One thing is all."

Here was Dean as Taoist sage, teaching the oneness of all rather than the intricacies of the half-court trap, a coach who said a lot by saying very little. Chapel Hill had always been a sucker for blending the wisdom of the East with its own laid-back predilections.

"The Day" was just the first recitation of Mike Troy's lunch hour. Every few minutes or so, he would break into another. He had poems for all occasions. Broad-browed, shorts-wearing, muscular-legged well into his sixties, and lugging a backpack stuffed with verse and clippings, he segued from ribald jokes to local history to spontaneous orations with such uniformity of manner that it was hard to tell what was what.

Two older couples dining at a nearby table kept looking over at us. Their distress may have been provoked by Troy's loud repetition of the golfer Sam Snead's advice for a long and vigorous life: "Always fuck a different woman every day."

Mike Troy had lived most of his adult life in Chapel Hill and was the kind of guy you rarely bumped into anymore in this modern era (in fact, he'd recently moved to Hillsborough). Chapel Hill was so chic and professional now, but once upon a time there had been a loose tribe—let's call them the Iconoclasts—a class for whom perpetual studenthood constituted the ideal life. Slackers before the term existed, these fellows (they were almost all men) constituted the last survivors of the golden age of Chapel Hill lassitude, a period that ran from the beatnik Fifties to the hallucinogen-friendly Seventies. For decades, they had wandered Franklin Street at all hours of the day and night, clustering at Jeff's or Danziger's or the Tempo Room (gone now, all of them) or on the steps of the post office, ready to unleash their impromptu philosophizing on unwitting passersby, just the way those old Greek fellows did it back in Athens a couple thousand years before. They worked in record shops or bookstores, libraries or restaurants, out-of-the-way realms that allowed them the freedom to think whatever they pleased. Mike Troy, trained as a lawyer, had carved out an interesting niche for himself as the proprietor of a bar called He's Not Here, in which I had spent many a college night imitating the great Irish writers, at least the part of their lives that involved not writing.

"Did you ever practice law?" I asked him.

"I practice now."

"What kind of law?" I hadn't been aware of this.

"I started practicing philosophical and poetical law. I try to help people decide how to do what's right."

To follow Mike Troy's conversation, you had to abdicate your own destinations and ride along in the rickety truck of his monologue as he drove from one obsession to another. But all of this swerving was part of Mike Troy's appeal. Plus, his heart was in the right place. Two places, actually. He loved both Duke and North Carolina, having discovered like an alchemist in a secret laboratory how to make gold out of the union of cultural opposites.

This was very interesting, very odd indeed. But then, so was Mike Troy.

And he was interesting and odd in a way that seemed emblematic of Chapel Hill, at least the way the town liked to indulge a view of itself as a zoo of tolerance, in which the keepers mingled with the animals until you couldn't remember who was an animal and who was a keeper and everybody was very proud of this confusion.

It wasn't just Dean Smith that Mike Troy commemorated, mind you. Back in 1992, after Duke won the second of its back-to-back championships, he wrote a poem called "The Game," dedicated to Coach K and his team, and which he now rousingly recited. The last verse said this:

> *There's a miracle waiting*
> *In each human heart*
> *Slap the floor all together . . .*
> *Let the miracle start.*

"I think I prefer the first poem, about Dean Smith," I said. It was beyond my capacity as literary critic to appreciate a paean to Duke floor slapping. Walt Whitman could have written about floor slapping and I wouldn't have liked it. Emily Dickinson could have scrawled an enigmatic couplet about floor slapping and that also would not have been okay. I lacked sufficient scholarly objectivity and I told Mike Troy that.

"In 1992, when Duke won the second championship," he responded, "that was the first time in my life when it became safe and even popular to wear a Duke T-shirt on Franklin Street, because everybody in the world goes with what works today. Before that you could get your head pounded in."

It had certainly been observed that everybody loved a winner. This was true even in the South, which now only infrequently brandished its defeat in "the war" as a badge of honorary difference from the rest of the United States. These days, winning tended to wash away whatever sins or reservations had been attached to the winner. And maybe that was what Troy meant in saying that everybody in the world goes with what works today. This was not a philosophy that I much appreciated.

Anyway, it turned out Troy *had* gotten his head pounded in. His mouth, anyway. "I can still feel the cut in my mouth where a guy punched me out when I was at Duke in law school and came over to Chapel Hill one weekend."

"I'm sorry for your pain and humiliation," I told him, "but hearing about this makes me feel better about your Duke poem. What happened?"

"I had been dating this beautiful girl from Greenville named Ruth Young. She was a great shagger, and I always wanted to go with her."

"That's the only dance I don't have to fake," I told him. "The shag."

"So she came over and spent the weekend at my parents' house in Durham and I took her over to Chapel Hill and this little bitty guy said, 'Well, Duke ain't shit.' I said, 'Well, we'll see about that.' I took off my cardigan sweater and gave it to somebody and I went looking for the little bitty guy. And that's when this great big guy hit me right here in the mouth." He pointed at the particular section of his mouth where the injury had been done. "Back then, if you couldn't kiss a girl, you couldn't do nothing. And that punch really messed up my whole plan with Ruth Young."

In the equality of his affections, Troy seemed to represent the possibility of synthesis (Duke = thesis, North Carolina = antithesis), the reconciliation of opposites, the union of the anima with the animus. He embodied a vision of the Peaceable Kingdom, in which the Blue Devils would lie down with the Tar Heels, in which Dean Smith and Mike Krzyzewski would embrace like sons of the same mother. As a certified hater, I could not imagine this enlightened realm. It seemed inhuman.

"How can you be both a Duke and a Carolina fan?" I asked.

And as Southerners often do (though less and less), he answered with a genealogy. "Back in 1925 to '26, my father was the last starting end on the Trinity football team," he said, "the first one on the Duke team." He and his brothers had all attended Duke, Mike Troy in the midfifties after graduating from Durham High School.

Ten years ago, he'd been reading a history of Trinity College when he discovered, much to his surprise, that in 1851 his great-grandfather had been a founding trustee of Union Normal College, which became Trinity in 1859. Troy's ancestors had traveled from Randolph County to Durham when Trinity changed locations in the 1890s.

"Trinity started off a lot poorer than Carolina ever was," he said.

He was at pains to establish how Duke, though private and fat with endowment now, and North Carolina, public and beseeching the state legislature for support, arose out of the same red clay, possessed a common ancestry, and were sponsored by the same angels. "I've always said there needed to be greater cross-pollination between Carolina and Duke," he argued. The two couples heard the word "pollination" and glared at their plates, shaking their heads in disgust.

Troy didn't notice. He went right on talking, as he was wont to do. "And if there were this pollination, then Duke would be the greater beneficiary. Because Duke *is* more elitist. Everybody is from New Jersey and has a furrow in their brow and looks pissed off all the time. That's what Franklin Street has already been for: for Duke students to come over and just get *unpissed*-off."

"Money follows honey," he said loudly, concluding that line of thought. He swerved onto a slightly different road, following his own tire ruts. "You know, old Horace Williams argued by antinomies," he said. "He would find the contradiction between two principles, each taken to be true and opposed. And I am doing the same with Duke and North Carolina."

It was time for another round of verse. "I did some rap songs," Troy said. Uh oh. I was getting the hang of Mike's poetry but the idea of listening to a white guy in his sixties rock the mike was alarming.

"You remember this big old black boy from Burlington, used to play basketball?" Troy asked me.

"Geoff Crompton?" One of Dean Smith's recruits from the mid-Seventies, Crompton had never lived up to the promise he showed as a high school player. For many years until his premature death, he strode around town in a kind of modified hippie garb, a leather pouch dangling from his waist. He was a sweet fellow.

"I would sit on Geoff's knee and he would put his hands up like I was his puppet and we would tell midget jokes. One time, I wrote a rap song and I had Geoff perform my doo-dah line. Everybody said I should record it."

He started rapping, nodding his bald head with the beat. The two old couples stood up to leave. On second thought, I liked the idea of Mike

Troy rapping. Let's serenade that censorious old foursome's departure
with a couple of lines, shall we?

> *I've got my paper and I've got my ribbon*
> *I'm telling the truth and I ain't fibbing.*

MANHATTANHEEL

The morning after I saw Mike Troy, I logged on to the website *Inside Car-*
olina for the first of several daily visits. It was the place where you could
always go where everyone knew your fake name. Coffee in hand, I
clicked on one thread after another, a hunter-gatherer in search of big
games. I liked screeds, jeremiads, vindictive attacks. I liked shoot-outs
between posters. I liked accounts of pickup games, recruiting gossip,
and the never-ending assault on Duke University. If I had ever thought
myself consumed in a most unseemly fashion with the Duke–North
Carolina rivalry, I had finally found a place that made me feel like a dull
elder statesman who parsed his sentences carefully and made a career
out of avoiding controversy. The guys here wielded flamethrowers.

The prototype for the site had been built in 1996 by Ben Sherman,
then a 16-year-old Carolina fan in the unlikely outpost of Newtonville,
Massachusetts. He couldn't quite say what resulted in his having a crush
on a team eight hundred miles south, but it emerged at about the same
time as consciousness. "I hated Danny Ferry as soon as I started watching TV," he said. Naturally, he applied to North Carolina for college,
where, as was so often the case with out-of-state applicants, even ones
with good grades and high test scores, Sherman was rejected. He headed
south anyway, to the University of Richmond, where in August 1998 he
established a new base of operations for his Carolina website. The site
had since evolved into a sort of vast cybernetic encampment, in which
hardcore Tar Heel fanatics could gather to express their tribal affiliations.

"The way I look at it, it's a grander scale of when me and fellow Carolina fans get together, just screaming and yelling and throwing things.
After the game, if you're pissed off . . . you vent. And a message board is

a place to vent. It's so ecstatic after the win, so devastated after the loss. Sometimes it boils down to the chance for somebody to say: I hate this."

The Duke–Carolina rivalry burned perpetually on the *Inside Carolina* board like one of those underground coalfield fires in western Pennsylvania. There was no way to put out the flames. Hardly a day went by without two or three new Duke threads sparking up, so many that some posters complained about the board's obsession with the Blue Devils. It wasn't that the complainants wished to defend Duke—to the contrary. Instead, they felt that such an obsession was demeaning to North Carolina. Such threads ceded too much psychic territory to Duke. They actually enlarged the school by spending so much time cutting it down to size.

In the hours I spent each day rummaging around the site, I'd gotten to know the personae of many posters. Ironically, one I'd come to appreciate was a Duke supporter, Lpark, everyone's favorite Blue Devil for his generally balanced and occasionally self-deprecating commentary on both programs. What was he was doing so far away from home, spending hours a day on the Carolina message board, suffering the jibes and taunts of his declared enemies? I couldn't say. Penance, maybe.

But from time to time he appeared just sane enough that the flaming partisans of *Inside Carolina* envisioned a potential convert. They were always trying to lure Lpark into the light—the light blue, anyway. They wanted to see him healed of his Duke affliction, purged of the demons that must have lived inside of him. Occasionally, Lpark got mad and punched back, as he did in one exchange: "You talk about Dahntay's lip snarl," he wrote, referring to the much-hated Blue Devil Dahntay Jones, now playing (rarely) for the Memphis Grizzlies. "How about Jerry Stackhouse's head bobs, Rashad McCants's primal screams, every face Vince Carter made after a dunk. What's the difference? The answer . . . the shade of blue."

This was entirely too reasonable a response, and thus doubly annoying. He accused the IC regulars (who tended to descend on Lpark for an old-fashioned street mugging when he got a little too uppity), of "taking any situation [in regard to Duke] and tagging it with the most sinister spin possible." Well, of course. He was posting on a North Carolina website, was he not? Go back among your own, he was instructed at such times. You want balance, you want nonpartisanship, join the League of Women Voters.

His polar opposite in sectarian terms (although equally possessed of intelligence, long memory, and the capacity for ripping new assholes in irksome posters) was the poster known as ManhattanHeel. She displayed a saucy, take-no-shit manner. One of the most visible and respected (and even feared) of the site's posters, Manhattan, or MH as she was often called, reminded me of James Carville and George Stephanopoulos, Bill Clinton's campaign team in 1992. Post an opinion about Duke, especially one that admitted even a speck of favor, and she arrived on the scene instantly with a definitive rebuttal, a one-woman quick-response team.

Here, for instance, is a post of Manhattan's occasioned by Donald Trump's visit to Cameron for the second Duke–North Carolina game of 2004. It serves as a fine example of the withering, take-no-prisoners tone Manhattan deployed against the Duke universe, or anyone who so much as sat cheering in a ringside seat at Cameron Indoor Stadium.

The only man I hate in America as much as K . . . is Donald Trump. So how fitting is it that he "just loves" Coach K? Let's see, Trump is egomaniacal, narcissistic, win-at-all-costs, an adulterer, a philanthropist for PR purposes only, and often a liar. He and K must have been twins separated at birth. The only difference is that Trump gets former Miss Universes and K gets Mickie.

Now, this is not just an uninformed rant. I have met Donald Trump. We share a degree from the same business school. I have a good friend who works directly for him (and no, a cheesy TV show was not involved in his employment). I have heard the untold stories, both business and personal. Basically, this is a man who would put his name on every roll of toilet paper in America if he could. You see the name TRUMP tackily plastered on as many buildings in NYC and Atlantic City as you see K's name blanketing every surface of Durham County. This is also a man who has repeatedly screwed bondholders out of hundreds of millions of dollars when a deal doesn't go his way. I'm sure he will be a star lecturer at the K Institute of Ethics.

So I ask you, is there anyone in America who is a better "face" for the K fan club than The (Other) Donald? Seriously, if I were a dook alum, I would be embarrassed to have yet another goon like him shilling for dook u. Just when I think it can't get any more ridiculous over at that faux gothic cesspool, it does.

I drank deeply from Manhattan's bottomless pool of vitriol. There was something tonic about it. On *Inside Carolina*, she was merely one among an entire nation of raging beasts, all as anonymous as members of the witness-protection program. And, oh reader, what a harsh world theirs was, full of fear and loathing. But at least that tiresome Carolina rectitude disappeared, replaced by an honest expression of dark, human passion that Nietzsche would have been proud to witness. Whenever the occasional wet blanket of a poster tried to plump for a little civility (there actually used to be a board member named PreacherJohn, or something like that), he or she was chased off.

That's because the *Inside Carolina* message board, unlike the state of Virginia, was not for lovers. It was for haters. All right, it was for lovers and haters. But the haters were the most enlivening. (The same was true, by the way, of Mike Hemmerich's *Duke Basketball Report*. When that site tried to be high-minded, as it occasionally did, it came off as too Olympian, even condescending. When it stooped to conquer, and let the nastiness rip, the *DBR* was enlivening, even for a Carolina fan.)

All of this suggested why ManhattanHeel was such a star. Her writing—reliably tart, deeply invested in the subject, and suffused with a long, poisonous memory, especially in regard to all matters Duke—charged the board with the vicious electricity of partisanship. Like American political culture these days, where being right—or left—meant never having to say you were sorry, sports websites thrived on polarity, especially when it was flamingly theatrical.

As a sassy gal in a guy's world, Manhattan attracted a lot of attention. My pulse raced a little faster whenever she landed on a thread. She was all that and she hated Duke. I decided I had to meet her—for sociological reasons, of course. And so with the assistance of Ben Sherman, I made contact.

Just from her posts, I had gleaned a little bit of her history. She'd lived for a while in New York while employed by some sort of financial services company; hence her board handle. But she had actually grown up down south and had returned home. She was married, and she lived in Charlotte, North Carolina's largest city, where she worked in real estate capital markets for the Bank of America. One Saturday, I went to see her.

We met for an early lunch on a Saturday morning so we could catch up later with her husband, Gary, and watch a game early that afternoon. The Mexican restaurant we had chosen was empty of customers. I asked her how much time she spent on *Inside Carolina*, and she told me she kept it open on her work computer "pretty much all day." "I just go over there, check it out, post a few things, and then go back to work." She grinned. "Multitasking."

Before a waitress had even showed up to take our drink orders, we had trashed Dick Vitale. "He and all the commentators make Duke out like it's an Ivy League school, the way they handle their athletes," she said. "And we all know it's not. And the kids that are going to Duke, they're not North Carolinians. They get there and basketball becomes their social life. They're more impressed with what they think they contribute to the game as Krzyzewski's 'sixth man' than with what is actually happening on the court."

Her interactions with Duke graduates had only reinforced her perception of the school as a magnet for snobs-in-the-making. "I've just found all of them to be pretty smarmy," she said. "Maybe I'm guilty of a preconceived notion, but I'll give you an example. After finishing at Wharton, I went to Wall Street. I was working for what was then Paine Webber from 1997 to September of 2001. I'm a vice president of my group. We were always recruiting and interviewing associates, and once we got this kid whose father knew somebody. Someone came to my office, handed me a resume, and said, 'You need to interview this kid.' The first thing I see is that he was a Duke University grad. Honest to God, the first thing that ran through my head was that here are 30 minutes of my life that I'm never going to get back because I knew there was absolutely no chance I was going to hire a Duke kid. And I knew that before even meeting him. Now is that fair? No."

"True," I said. "But that's life."

"But then the kid comes in. And we actually got along pretty well. He had played soccer at Duke. And he was a smart kid. So it was going pretty well. And I actually said in my head, 'Oh, my God, this guy's pretty good'."

"That was really messing with your worldview."

"It was causing me a conflict," she said. "So we kept talking. And he wasn't from North Carolina. Shocker. He was from—not New Jersey—Virginia or somewhere. He said something about North Carolina. I said I actually grew up there. He said, 'Oh, did you go to school there?' I said, 'Yeah. I went to UNC.' He goes, 'I'm sorry.'"

"Uh-oh."

"I know he felt like he was trying to be funny. But I'm like, okay, dude . . ."

"This interview is over."

"You're a kid interviewing with the vice president of a group and you insult her. Before I even met him, I knew I wasn't going to like him and guess what happened: He confirmed that."

The Duke basketball players struck her similarly—as monsters of striving and entitlement. "I was at Carolina when Bobby Hurley was playing at Duke. I couldn't stand him. Those guys are always such media darlings, too."

"Like Wojo," I said. The name immediately sprang to mind. How could it not? Steve Wojciechowski stood in for every obnoxious, overachieving white point guard who ever played the game. In order to show the coach how psychotically into playing he was, Wojo was the kind of guy who ran to every huddle like Nutty Buddies were being handed out there by the Good Humor ice cream man. He was the sort of lead-footed guard who commentators like Dook Vitale were always saying made up for lack of native talent with hard work. Vitale and his media brethren shilled Wojciechowski right into being named National Defensive Player of the Year for 1998.

The truth was that Wojo played with such effort and intensity largely because he was so slow! It required vast expenditures of energy for him just to catch up to the play and the man he was guarding. Wojo slapped his defender around like a girl in a catfight. *You bitch*! After one mid-Nineties Carolina–Duke game, the North Carolina point guard Ed Cota displayed his scratched-up arms to the press and tried to explain Wojo's defensive prowess, saying something to the effect of "he fouls. A lot."

His style was so antithetical to physical grace that he made you think about inequalities of scale, of the fundamental unfairness of the universe when it came to the distribution of gifts, in this case physical talent. Worse, he made you wonder why white players tended to be perceived as hard workers and black players as naturally gifted. He made you see the lingering taints of race when all you wanted to do was watch a basketball game. None of this was exactly his fault, either. But his desperate, hand-slapping, knee-to-the-thigh scrappiness still rankled.

"I hated Wojo," she said. "And now," she said, "JJ Redick is starting to be that way. My husband called it. We were watching Duke play Georgia Tech. JJ hit a shot. He started talking so much shit. He was mouthing off while he was running backward. Gary's like, Holy shit, look at Redick talk. He's just like the rest of them. And you know what I say? It's like their birthright. They think that they've got Duke on their chests so they can act how they want. That comes from their coach.

"I don't know if this is like Carolina fans just being delusional, but I really think that our program is run very differently. I'll use as an example when Matt Doherty is in the huddle at Cameron and he makes a joke about the Duke cheerleaders being ugly. He apologizes publicly for that. Then you look at things that happen in K's program that are 80 times worse than that and it's like—" She turned her palms upward and shrugged.

"In the end, it's the sanctimony," she said. "Trying to seem better than they are. That's the reason that Duke hate is nationwide now. It's not just Carolina fans. I guess the Dukies can say they're hated because they've won so much, but have they really won so much? In the ACC I can understand that view, but they have only one national championship since 1992. But the Dukies like that they're hated. It validates them."

REASON IS THE PRESS SECRETARY FOR THE EMOTIONS

And here at last one of the fundamental principles of Duke–Carolina hatred was revealed to me in its bluntest form: First you love North Carolina, then you hate Duke. Love prefigures hatred. In fact, love necessitates hatred. And the rationale for that hatred follows only in the wake of its existence.

There is a psychologist at the University of Virginia named Jonathan Haidt whose research suggests this sort of thing is the norm—that humans judge, *then* they reason. "Reason," he asserts, "is the press secretary of the emotions."

Haidt conducted a variety of experiments designed to reveal the actual way we go about reaching our moral judgments. One involved the presentation of the following scenario to the test subjects. A brother and sister, Mark and Julie, are vacationing in the south of France. They drink a little wine, and as an interview with Haidt published in *The Believer* magazine puts it, one thing leads to another, and they decide they want to have sex. The couple uses two different kinds of contraception, enjoy their experience, but decide they will never do it again. Were they wrong to do what they did?

The subjects of Haidt's experiments generally reacted to the scenario by stating emphatically that the brother and sister were indeed wrong. But the reasons they propounded were usually subverted by the story (e.g., the couple used two different kinds of birth control, so the likelihood of a baby being born with genetic defects is highly unlikely). In the end, the experiment's subjects reached a state Haidt calls "moral dumbfounding." All of the standard rationales for faulting the couple were stripped away from them. They sensed that something was still wrong with the couple's actions, but they had no words to explain why.

The experiments point to the powerful role played by intuition and emotion in our ethical stands. In essence, our moral views arrive before our justifications for them, by overnight delivery, as it were, with our justifications arriving by standard mail two or three days later.

TWELVE

The Endgame

PLATO'S SYMPOSIUM

The ACC Tournament

MCI Center

Washington, D.C.

March 10–13, 2005

HERE I WAS, a middle-aged man going out the door on the way to Washington, D.C., where I had arranged to meet an erudite Tar Heel fan named Thad Williamson at the ACC Tournament, so that we could catch North Carolina's game with Clemson. For years, I had been reading his evenhanded dispatches about games on *Inside Carolina*.

"Do you need any money?" my mother asked.

"About fifty thousand dollars," I said.

"I can give you twenty," my mother said.

"I'll take a rain check," I said, kissing her good-bye. "But I appreciate it."

A few hours later, Williamson ambled up to me outside the MCI Center on the edge of Chinatown. In his midthirties, lanky, bearded, wearing boots and jeans, he looked the part of a young lefty academic, which he was, but in his eyes, I discerned the anxious, hot-wired glint of true fandom. He sold me a ticket for face value, pocketing the cash so that he could take his sister and brother-in-law out for supper later. We blundered around, looking for an open Chinese restaurant in the shadow of the MCI Center. We ended up at the one at which we had started, the only diners, hunched in a black leather booth, enduring a flyover by lacquered ducks.

Williamson had written a book called *More Than a Game: Why North Carolina Basketball Means So Much to So Many,* probably the only sports book likely to ever be published by Economic Affairs Bureau. It's an encyclopedic tome, 320 pages, including a statistical analysis of the beliefs and habits of more than six hundred UNC basketball fans who answered the author's extensive questionnaire. Although Williamson's prose is clean and sober, as befits a book that aspires to sociology, *More Than a Game* is nothing less than the political autobiography of a young scholar whose ethical development, like so many others', has been profoundly influenced by Tar Heel basketball, and in particular Dean Smith, as an exemplar of the good life.

Of course, it is hard to imagine another successful American coach having Smith's moral suasion. Bear Bryant had a hound'stooth hat and that great saying to counter excessive celebrations in the end zone: "Act like you've been there before." But it took his team being pummeled by Southern Cal and its black running back Sam Cunningham for Bryant to wake up on race. And his awakening—"Where do I get one of those guys?" he is said to have asked in reference to Cunningham—smacks of opportunism, not ethics. Vince Lombardi was a great coach of the Green Bay Packers, salty and hardworking, but his sentiment "Winning isn't everything, it's the only thing" is the sort of inane remark that suggests that total victory is the norm to which Americans should aspire in all endeavors.

The list goes on. Adolph Rupp of Kentucky: great coach, regrettable bigot. Phil Jackson: terrific tactician with a repertoire of groovy little Zen parables, but also strangely receptive to the blandishments of Hollywood, to the lure of celebrity, and to the pseudo-wisdom of whatever New Age nostrums washed up on the beach that day at Malibu.

And of course, there is the yin to Smith's yang, the no to his yes, the

South to his North Pole, Mike Krzyzewski. The Duke coach's lust for victory often appeared to come at the expense of compassion and decorum. And his influence seemed limited to the standard penumbra of allure that the successful always radiate. *Be like me and win more, make more money, kick more ass.* It made sense that his symbol for team unity was a fist. Businessmen might want to co-opt some of his motivational strategies (publishers apparently believing that middle managers have the same chance to "win" at their jobs as coaches), but it would be hard to see how his example would lead to anything more noble than the aggrandizement of self.

As we raced through our meal, Williamson, now an assistant professor of leadership studies at the University of Richmond, gave a quick account of his pilgrim's progress. His conversion experience arrived one night in 1974 when he was four or so years old. His parents were departing to Reynolds Coliseum in Raleigh to watch North Carolina play NC State. "I want North Carolina State to win," Thad told his brother George. As is customary in tales of redemption and enlightenment, a compassionate elder set the sinner straight. That elder was George, age nine or so.

"No, that's wrong," George told Thad. "We root for UNC. That's who we are, and that's who you have to root for. You actually want UNC to win, not State."

Williams's response: "And so it was I became a Carolina fan. George's lesson sunk in immediately." Young George is Socrates to young Thad's Plato. In his book, Williamson dresses up the moment in the stiff muslin of political philosophy: "George also imparted a valuable and lasting lesson about the value of particularity ('we all root for UNC') as opposed to a diffuse wish for the aggregate common good (the notion that the 'state' would be somehow more important than just 'the university'). Both attachments to particularity and a general wish for the greater good are important human values, important habits of thought and living. But only one is capable of constituting an individual's identity in a deep, inescapable sense, in a way that shapes your very perception of reality and touches the deepest well-springs of the heart."

Take it to the hoop, Plato.

Because his father was a popular teacher at the university and actively involved in university affairs, Williamson enjoyed the occasional chance to hang around the team, sometimes traveling to far-away games. He knew the assistant coach Bill Guthridge from church. As a 12-year-old in 1982—that miraculous year!—Williamson landed a prized position at Carmichael Auditorium during games as one of the kids who flipped the antiquated wooden panels of a scoreboard that the UNC coaches still used because it showed up better in their film. He kept the job through high school.

Then Williamson departed Chapel Hill for Brown University in Providence, Rhode Island, where along with politics and sociology, he also learned how to accentuate his Southern accent to heighten his visibility among the female species. Most of his activist buddies there didn't quite understand his continuing obsession with Carolina basketball, but that didn't deter him. He befriended students who had cable, flew home occasionally for games. In 1995, he began to cover the Tar Heels for an early version of *Inside Carolina*, and continued to do so while studying for a master's in Christian ethics at Union Theological Seminary in New York and then while taking his doctorate in political science at Harvard.

Williamson proposed to his girlfriend, Adria, in the wake of the couple's trip to the Great Alaska Shootout in 1997. At the Seattle–Tacoma airport on their way back to their homes on respective coasts, they lingered, unsure of their course after four and a half years of dating. "Well, we could get engaged," Adria ventured. And they did, right there and then.

Looking back on it, Williamson writes, "I'm sure there would have been at least some buzz between us after our great trip to Alaska *if Carolina had lost* [italics mine]. . . . I'm not sure whether, if the Alaska tournament had turned out differently, what happened an evening later in the Seattle airport would have been different as well. My wife and I are both glad we didn't have to find out." In other words, the likelihood of this man's marriage rested on a succession of Carolina victories!

How well I understood this, the way a Tar Heel win could surge into you like the Holy Spirit, enrapturing and enlivening you, making you capable of deeds of valor you wouldn't have otherwise attempted. I gave the woman who would one day become my ex-wife a diamond ring on the pier at Sunset Beach, North Carolina, and we then hurried back to our motel room to watch the Tar Heels lose to Syracuse in the 1987 NCAA regional finals. I didn't take it as an omen at the time. At the time, I thought I had simply underrated Derrick Coleman and Rony Seikaly.

As a writer and thinker concerned with the greater good, Williamson required of himself a strenuous moral accounting. Somewhere along the way it occurred to him that his fandom ought not to be fanaticism. And so he went to Duke's Cameron Indoor Stadium one night to test himself under the guise of doing a little anthropological fieldwork. He returned to write what will probably remain the most controversial piece of his scholarly career, "The Virtues of Duke Basketball," published in 1999 in *Inside Carolina*. It probably also stands as the only essay about the Duke–Carolina rivalry to cite Emile Durkheim, the founder of modern sociology.

"It received far more response than any other thing I've written," he said. A lot of UNC partisans had taken issue with Williamson's evenhandedness. "Most Carolina fans think of their own program as a class act, committed to basketball, yes, but in the right way, a way that expresses a sort of virtue," he had written. "It's hard to see Duke in the same light, but high time to do so." If Duke went on to win the national championship that year, he explained (they eventually lost in the finals to Connecticut), "there may be a little jealousy but no real grudge from this corner."

"You're a brave man—and saintly," I said as we shoved aside our plates and called for the check. My fortune cookie read: "You are tasting the sweets of success." Not yet, I wasn't. If there was one thing I hated (and there were clearly many), it was a fortune cookie that spoke too soon. As it turned out, I was right to be wary.

Together we hiked up the stairs of the MCI Center and settled into our seats next to a couple of Wake Forest coeds wearing "Free Chris Paul" T-shirts. The Wake point guard had been suspended a game for punching North Carolina State's Julius Hodge in the nuts, occasioning perhaps the first public display of sympathy in history among fans of other ACC schools for the trash-talking Hodge. "Chris should be suspended," one of the girls said, "but he's not a bad guy."

I was hoping that the Clemson game would be one of those relaxing, quarter-final affairs—a few dunks, a big lead, a little sweat—the basketball equivalent of watching traffic flow by on the highway on a summer afternoon.

And at first, that's what it looked like we would get. Raymond Felton opened the proceedings, hitting a long three. "Not a good shot," Williamson said. "But we'll take it." He had that longtime Carolina follower's appreciation for what constituted the right play, and the suspicion of the wrong play even if it had gone right. He knew that it boded poorly for the future.

Initially, the crowd was becalmed. It was too early for basketball. A North Carolina State fan behind us yelled in protest: "No replays, no beer!"

The Tar Heels went out to a quick 6-to-0 lead. Then Melvin Scott missed a three from the corner. He was starting yet another game in place of Rashad McCants, who was nonetheless suited up today and scheduled to see action. "A little rust," Williamson said.

What had rusted the most was the North Carolina defense. The players seemed creaky, in need of oil. Clemson would penetrate and kick out for threes and the Tar Heel defenders would arrive a step late. Shawan Robinson was having a field day from the outside. Or the Tigers would keep the ball and attack the basket. With only 14 seconds in the period, the Tigers nailed another three to lead at the half, 43 to 40. The largely anybody-but-Carolina crowd erupted with glee. The UNC defense wasn't just rusty; it was shot full of holes.

The second half was more of the same.

With 15 minutes left in the half, Clemson led by six. The Wake Forest and North Carolina State fans were roaring now. "Those fans shouldn't invest so much emotional energy in this game," Williamson said. "They're only setting themselves up. Now if it's Clemson up five or so with five minutes left . . ." He left the thought unfinished. Neither one of us believed such a scenario likely. But we should have. Because, in fact, with only four minutes to go, Clemson was ahead by five at 77 to 72, having led by as much as 13. With just over two minutes, the Tigers still clung to a two-point margin.

Were it not for a single-handed rescue by Raymond Felton, who was on his way to a career high of 29 points, the game would have been lost ignominiously, the first time the number one seed had lost to the number nine seed. He hit a three with 1:21 left, to give the Tar Heels their first lead in the second half at 81 to 79. The fans around us went stone quiet. "I told these people not to get invested," Williamson said. "If we win, I've got to go to the concourse and holler. It's just something I do. So don't be alarmed."

"I'll join you," I said.

When the game was finally secured at 88 to 81 after the final two of Felton's 11 free throws, Williamson uncoiled his lengthy frame from the seat, meandered to the teeming concourse, cupped his hands around his mouth, and loudly screamed, "WHOOOOOOOOO!" A smile creased his face for the first time in an hour or so.

I will say that for an objective journalist, my scream wasn't half bad, either. The two Wake coeds in the "Free Chris Paul" T-shirts happened to be passing by at that moment and they looked concerned. This made the beast and me very happy indeed, which along with the young scholar Williamson made at least three of us who sauntered along the concourse in fine fettle.

Unfortunately, the rest of the weekend provided no further comebacks. The next afternoon, Carolina played desultorily and lost to an energized Georgia Tech. Duke, meanwhile, beat NC State and on Sunday, in the championship game, held off the Yellow Jackets, 69 to 64. I drove back to Chapel Hill in a despondent slough and recovered my vim only when I saw that North Carolina, like Duke, had been awarded a number one seed in the upcoming NCAA tournament.

STEP UP AND BUST IN THEIR EYE
NCAA Regionals
March 25–27, 2005
Syracuse, New York

At one point this season, the posters on *Inside Carolina* were discussing their favorite Melvin Scott moment. One wrote that his was how "after the NC State game this year when Melvin started, he said something afterward like 'I don't want to toot my own horn, but toot toot!' Another poster delivered a quote of Melvin's that had acquired legendary status on the board as all-purpose exhortation: "Step up and bust in their eye."

HickoryHeel01 had written that "it's almost a cliché to say 'the best is yet to come,' but I believe that could be the case with Melvin. There will come a game in the postseason where Melvin hits the big shot or shots to win the game."

The chances to do this were dwindling, however. In Charlotte, the Tar Heels had raced past Oakland and Iowa State during the first two rounds of the NCAAs, seemingly recovered from their loss to Georgia Tech during the ACC Tournament semifinals. At the same site, the Blue Devils had bested Delaware State and Mississippi State.

Tonight, Duke and North Carolina were both playing in the Sweet Sixteen, the Blue Devils at seven, the Tar Heels at ten. The evening began auspiciously. That is to say, Duke was in big trouble in its game with Michigan State, a development about which the beast was hollering his horned head off and that the journalist found more than palatable. It had been a month since my flirtation with the dark-blue side and without the example of actual humanity in front of me to complicate my bias, I was reverting to my usual foulmouthed self. This was also attributable to watching the games on television, an invention that by flattening and melodramatizing people had induced more mob behavior than any cracker with a firebrand ever had.

At the half, the score was deadlocked 32 to 32. Shelden Williams and Daniel Ewing were giving Duke a nice inside-outside balance. Led by its center, Paul Davis, Michigan State was commanding the boards, on the way to 16 offensive rebounds for the game. But in the first half, the Spartans missed every three they attempted, going zero for six.

Not in the second. Then they nailed five of six and took tenuous control of the game with a 42-to-36 advantage after back-to-back threes. The Blue Devils fought hard but could never recapture the lead. With 27.4 seconds and Michigan State ahead 76 to 68, Duke's death agonies began. The phone began to ring. Mike C., my college roommate, called from New York. "I've been contributing in every way I can," he said.

"The key chain?" I asked.

At that moment, JJ Redick missed a wide-open three. He had shot only four of 14 for the game; the Spartans had defended him well. With that, the season ended for the Blue Devils. "See what I mean?" Mike cried.

For just an instant, the finality of the defeat shocked and tenderized me. It was like the suddenness of death. I felt bad for Redick; I imagined that he would be seeking the wintry consolations of *Nebraska* later tonight. But Mike's inarguable logic put me back on the straight and narrow. "I could deal with our losing the next game," he explained, "but not if Duke had won."

As soon as we hung up, the phone rang again. It was my mother calling from up north, where she was visiting my brother and his family. "This is Shelden Williams," she said, pretending to be the Duke center.

"It's all over for you, Shelden," I said. "Back to Durham you go."

Oh, we were a tough crowd, heartless, simply heartless.

But from here on out, everything would be gravy. Duke was going home.

Usually, the payback for taking pleasure in the misfortunes of others takes a while to arrive. But not tonight. In the next game, the Villanova Wildcats shot to a 5-to-0 lead over North Carolina and kept right on going. Having started four guards, they were the rare team that could actually surpass the Tar Heels' speed and quickness. And they were playing with the angry intensity of underdogs.

"God, or whatever You are," I prayed, "I know hatred is wrong. Is it too late to make up for my malicious, not to mention deeply satisfying, glee at the results of that last game? Just tell me what to do. And then, as a reward for my repentance, could you let North Carolina come back and stomp Villanova? It wouldn't have to be a stomp, though isn't it better to be straight with a team that's not going to win rather than give it false hope? So a stomp would probably be more ethical. But that's for You to decide. I can't think of anything else I really need at this time. I'll check in after the game."

The Wildcats, however, continued to pour it on, taking a 12-point lead within the first nine minutes. By halftime, North Carolina knew it was in a war. Four of its top players, including Sean May and Marvin Williams, had incurred two fouls each. Rashad McCants had hit one of six from the field, Raymond Felton one of eight. Felton was also on his way to six turnovers, the team to 17.

In the second half, Carolina recovered from the shock of Villanova's initial onslaught and tiptoed to its biggest lead of the game at 50 to 45. But before I could relax into the likelihood of victory, Felton picked up his fourth foul. Then Quentin Thomas came in and was immediately called for charging. The Tar Heels were on the verge of falling apart.

In came Melvin and within seconds he had hacked the Wildcats' Kyle Lowry, who scored on the drive and subsequent free throw. But from then on, Melvin played a near flawless floor game, directing the offense and getting the ball to Rashad McCants, who had nine of the Tar Heels' next 11 points. "I told him I was going to give him the ball," Melvin said, "and he had to carry us." With 2:11 left and UNC up nine, Felton returned to the game and promptly picked up his fifth foul.

This time, North Carolina panicked. With 40 seconds to go, Villanova trimmed the lead to a basket at 64 to 62. On the next sequence, Melvin, back in for Felton, was fouled with 28.9 seconds. He went to the free-throw line for the biggest shots of his career. As he told me later, he saw this as "a moment of clarity, a moment to prove that I could play at that level. It was a chance to end on a good note a season that had left a sour taste in my mouth." He had often envisioned a moment exactly like this, standing at the free-throw line in a hostile arena, ready to silence the crowd. What would Smitty say? Smitty would say, "Knock 'em down, young man." At last, his time had come. Melvin sighted the first. Swish. The second. Swish.

The game should have ended then, but it didn't. Instead, Villanova came down and Randy Foye got fouled taking a jump shot. He hit the first free throw, missed the second, but Villanova grabbed an offensive rebound. Allan Ray, Melvin's man, headed into the lane and threw up a runner that dropped in. At the same moment, the ref blew the whistle and it looked for all the world as if Melvin, who had been guarding Ray tightly, had been called for a foul. I groaned and put my head in my hands. With the basket and a free throw, the Wildcats would tie the game. Miraculously, instead of calling a foul on Melvin, the ref waved off the basket and signaled traveling on Ray. Melvin screamed in jubilation, punched his fist into the air, and, while running back down the court, appeared to vault an invisible hurdle. The Tar Heels were going to win. With Felton on the bench, Melvin had run the team without committing a single turnover against the quicksilver Villanova guards. Had he not stepped up and busted in their eye, the Tar Heels would have been on the midnight plane back to Raleigh–Durham, just like the Blue Devils.

Instead they survived. And on Easter Sunday, they won again, beating Wisconsin 88 to 82. Now they were headed to the Final Four. And so was I.

Oh, and thank you, God. I swear I'll be a better man really soon.

YOU GONNA THROW THAT IN MY FACE?

A few days before the Tar Heels departed for the Final Four, Melvin drove to Men's Wearhouse in Durham to look for two new shirts and a tie to wear in St. Louis. He was wearing his electronic billboard of a belt, a blinking strip of lights that could be programmed to flash different messages. This afternoon, the belt kept pulsing "REG SEASON ACC CHAMPS."

At the store, Melvin ran into JJ Redick, who was hunting for a suit.

"You gonna throw that in my face," Redick said, studying Melvin's belt and laughing.

The two discussed old times, like when they played at the U.S. Olympic Festival in Colorado Springs as high school students. Redick remembered watching Melvin shooting on an adjacent court. "Who is that guy?" he wondered. Redick had thought Melvin was as good a shooter as he himself was.

"I enjoy watching you," Redick told Melvin.

"I wish I were in your shoes to get those shots," Melvin said.

"Yeah, but there's nothing like winning," Redick said.

THE ROAD TO THE FINAL FOUR
March 31, 2005–April 1, 2005

For me, the road to the Final Four was not a metaphorical highway but an actual one. I took Interstate 40 to Interstate 24 to Interstate 57 to Interstate 64, all of which would shoot me to the Hyatt Regency next to the old train station in downtown St. Louis, where I had managed to secure a room at the hotel where CBS staffers were staying.

It was a bleak day of travel until I finally got in tune with the hum of the road. How bleak? Well, in the mountains separating North Carolina from Tennessee, I scrawled with my right hand while steering with my left: "In middle age, this is how I feel. We're just meat on a planet, a slaughterhouse orbiting the sun." Hey, I was some fun date! Basketball, deliver me! Had I crashed while writing this, it would have been a form of poetic justice, the wreckage steaming down in some Appalachian hollow. But then I would have missed the games.

As I crossed out of Tennessee in the cool of that spring night, the airwaves were surreal, a collage of disconnection and paranoia. I listened to the Michael Savage radio show. This guy was so right wing he was mad at George W. Bush. "We're the thirty percent who are upset about Terri Schiavo," Savage was saying. "And we're the same thirty percent who put George Bush in office." He thought that Bush should have commandeered Schiavo, taken her out of the nursing home, and installed her—where? The White House? Savage asked whoever had sent him the two great bottles of wine and the chocolate truffles for his birthday to call in and identify themselves. Otherwise, he was going to have to smash the bottles of wine in the streets of San Fran Freako, as he put it. He blamed everything on the spread of worldwide socialism. I supposed the next thing he'd be lobbying for was a preemptive strike on Sweden. I found a classical station out of Murray State; trumpets blessedly blew me across Kentucky.

For the first few days of any trip, I was exquisitely sensitive to new locales. I was like my father's people in that. It wasn't that long ago that for the Blythes to travel a hundred or so miles to Virginia was like our voyaging into interstellar space. Of course, in the days before the interstate, travel was harder and not taken for granted the way it was now. In the days before television and the Internet, regions were naturally more distinctive. Virginians might as well have been the Blue Men of the Sahara. To travel to their native grounds was to scent the difference ("smell that air!" my grandmother used to command), to feel the slightly cooler weather come winter. In Southside Virginia they pronounced house as *hooose*, store as *sto'*, tomato as *tamahto*. These sorts of things were marvels of the exotic for a family that found everything it wanted in the world at home. Even as a smart-ass child growing up in a university town, I just couldn't get over that *tamahto* business.

The Kentucky poet, novelist, and farmer Wendell Berry has written that "love defines the difference between 'the global village,' which is a technological and a totalitarian ideal, directly suited to the purposes of centralized governments and corporations, and the Taoist village-as-globe, where the people live frugally and at peace, pleased with the good qualities of necessary things, so satisfied where they are that they live and die without visiting the next village, though they can hear its dogs bark and its roosters crow."

This feeling was the gift my father had worked to bequeath us.

Although it was tricky. For him, loving one thing seemed to require that you hate another, that you divide the world into two disproportionate pieces: the inherently local and familial, that which was known and loved, and the unknown, the foreignness that threatened the gentle people behind the boxwood hedge in their Carolina yard. This isn't to say my father wasn't curious or a good traveler. He was both. But for him, there were always two places in the world: home and everyplace else.

So as I neared St. Louis, at the edge of the plains, the beginning of the West, I felt slightly unmoored amidst this expanse of flatness and sky, and homesick for vegetation and slope, the pastel greens of a North Carolina spring complicating the modest hills.

Perhaps you can understand why for me, the expression "home team" was not just a figure of speech. Which was why when I reached St. Louis

early the next afternoon, I immediately went to the Edward Jones Dome
to cheer the Tar Heels at practice.

Nearly all of the estimated 31,500 fans who had shown up to watch the
open practices that Friday afternoon were orange-clad Illinois support-
ers. They booed mightily when North Carolina, which had been allotted
the practice slot immediately after the Illini, ran into the coliseum and
started doing laps of the court. But you could feel the respect of the Illini
boosters as well. They studied the North Carolina players intently, with
what appeared to me to be a measure of fear. In their wary eyes, it
seemed that the Tar Heels were blue bloods again, that an old family
dynasty had returned to claim its rightful place after a brief wobble in its
familiar fortunes.

Where Illinois had used their time for a glorified shoot-around, the
Tar Heels were now scrimmaging full-court. "Don't wait, don't wait!"
Roy Williams yelled, waving Raymond Felton forward. He didn't want
Felton loping around the perimeter. He wanted the ball advanced
towards the paint to put maximum pressure on the defense.

They scrimmaged so hard that I worried about their legs tomorrow
against Michigan State, their semifinal opponent. A high school coach
from East St. Louis who had brought his team of braided players to
watch, said: "Carolina is really going after it."

The practice ended with a rousing dunk-a-thon. David Noel and Marvin Williams both awed the crowd with impressive slams. Sean May
bobbled a few and laughed. Then it was time for Melvin to show his
stuff. He bounced the ball hard so that it soared towards the goal. And
as it was rising, he stood there, pulling his practice jersey over his head.
Now stripped to his ever-present wife-beater, he ran to catch up to the
ball, jumped high into the air, caught and crammed the ball in one con-
tinuous motion through the hoop. The fans roared. Melvin won the
loudest ovation of the afternoon.

He bowed in all four directions to the crowd, and saluted them. He
was treating this as if it were his one shining moment.

He was followed by Marvin Williams, executing a cartwheel and
jamming. It was the gangliest cartwheel I have ever witnessed, but it

was still a cartwheel, and how many six-nine guys can bring off a cartwheel at all?

On Melvin's next dunk attempt, he fell flat on his ass. Practice was over.

ABOVE THE RIM
NCAA Semifinals
St. Louis, Missouri
April 2, 2005

Pope John Paul II died today. Before the games began, the NCAA staged a brief tribute. The Pope's picture appeared on the video screens arrayed around the coliseum. We all engaged in a moment of silence. There seemed to be a lot of Midwestern Catholics here and when quiet time was over they clapped and waved pom-poms for the Pope as if he were about to be introduced at power forward for the Illini. Much to my delight, I could hear my father grumbling: *Papacy.*

If he and the Pope were now in the same place, if such a place existed at all, I imagined my father would have been teasing the Pope for purveying all that silly hoo-hah about transubstantiation, the Virgin Birth, and the healing waters of Lourdes. "Pope," he would have been saying. "Do you really think all that hocus-pocus is necessary? Isn't all that stuff just a light-and-sound show for the weak of heart? Doesn't it really just boil down to one thing—love?" And the Pope—if he wore his mantle lightly, or if not a mantle that hat of his that looked like a napkin at a fancy restaurant—the Pope would have had to laugh. If, that is, he and my father were in the same place. And if there was such a place.

As the season was nearing its end, I felt myself moving towards my father's skeptical love, or loving skepticism, his complicated orders of affection. Increasingly, I was seeing through his eyes.

I wondered: Why was the NCAA doing this tribute at a basketball game? Why mingle two great religions?

Then again, maybe I was a little sour tonight. I was sitting high above the court because I was the bearer of a Limited Access press pass, which

I soon discovered was like walking around with a big banner across my chest that read: NOT A REAL SPORTSWRITER (which was true, but still . . .) SO HIT ME, BEAT ME, AND PUT ME IN THE UPPER PRESS BOX SEVERAL MILES ABOVE THE FLOOR.

Down on the court, you witnessed screens laid with brute force, body against body, grunts and thumps and curses. The hand-to-hand combat of play underneath the basket. The hip butts and hand checks of jockeying on the perimeter. Up here, the view was as sanitized as that from the cockpit of a bomber. It was like watching the locker-room chalkboard come to life, diagrams in motion, sneakers squeaking rather than chalk.

In the first game, Illinois appeared to believe it was the team of destiny. They were whipping the strut out of Louisville, reducing Rick Pitino to rare silence on the sideline. Having made a remarkable 15-point comeback in the last four minutes of regulation against Arizona in the regional finals and having won by a point in overtime, the Illini probably felt invulnerable. The Reverend Roger Powell, the Illinois forward and a licensed Pentecostal minister, was going crazy in this game. Sitting next to me, a sportswriter from a small paper downstate (as they say here) kept exclaiming, "The Rev!" every time Powell hit a shot. Clearly, we had both been assigned to the partisan journalist section. On one miraculous play, the Rev missed from behind the three-point line and seemingly at the same moment, arrived at the rim for a resounding stuff. "*The Rev!*" the writer and I yelled together. Us Limited Access types had to stick together. All around us, the Michigan State fans were actually cheering for fellow Big Ten member Illinois in a way that would never happen in the perpetual feud that was Duke and North Carolina. As Dick Vitale would say, "Are you kidding me?"

The final: Illinois, 72; Louisville, 57.

I snuck down to the court while Michigan State and UNC were warming up. With 22 minutes to go before tip-off, Melvin was practicing free throws. Sean May was scanning the crowd. Most of the players lay on the floor, stretching themselves with rubbery black cords. They all looked a little nervous.

Bridget, Charles, and Will Scott were sitting in the front row. "This man offered me several thousand dollars for my ticket," Bridget said. "I told him, 'Are you crazy? I'm not selling this.'"

Michigan State had beaten Duke 78 to 68 in the regional semi-finals—a definite point in its favor—then taken down Kentucky in double overtime, 94 to 88. Under the irascible coaching of Tom Izzo, they played a rough, physical style of basketball, clawing for rebounds and bodying offensive players to the extent that they could get away. Athletic at nearly every position, they matched up well with North Carolina. Michigan State merely had to find a shooter or two for the evening—Maurice Ager? Shannon Brown?—and they would stand an excellent chance of winning. This was not a point in their favor.

The game turned out to be a bipolar extravaganza. In the first half, UNC fell behind early. Raymond Felton threw the ball over Sean May's head. Marvin Williams and Felton knocked the ball out of bounds going for a defensive rebound. On a fast break, May lost the ball. Something was missing from the Carolina attack and it was usually the ball. The Tar Heels were sloppy, out of rhythm.

The Spartans rode our guys further out on offense than they were accustomed. And they were moshing on the boards with wild abandon. They went out to a quick 8-to-4 lead before North Carolina experienced a resurgence that featured three dunks, by McCants, David Noel, and Jawad Williams. Were it not for Williams, UNC would have been in big trouble. He was nailing threes, rebounding, and playing great defense on Alan Anderson, who had been leading the Spartans in scoring during the NCAA tournament with an average of 14 points a game. For Williams, this marked a major turnaround; he'd been in a slump for the NCAAs and some of his minutes had been farmed out to the freshman forward Marvin Williams. Jawad had not complained. He was inevitably stoic.

North Carolina edged out to a brief, jittery lead at 22 to 20 when May hit only his second bucket of the half; it would not have been reassuring to know at that moment that it was also his last bucket of the half. Michigan State was outmuscling the Tar Heels, but that's what Tom Izzo teams tended to do to opponents. They played like longshoremen trying to fight their way to the bar for a drink on Saturday night. Paul Davis, the six-eleven center who had once confessed to his coach that he didn't love the game of basketball (Izzo had responded compassionately by saying that he wanted to kill Davis) was pounding the backboards on his way to 15 rebounds. This evening, Davis looked as if he sort of liked the sport, kind of, maybe.

With the score 25 to 25 on a Chris Hill three, Roy Williams started channeling Dean Smith and put on the floor an unorthodox lineup of Melvin, May, Jawad, Noel, and Quentin Thomas. His teammates believed that QT, as Thomas was known, was a considerable talent at the point. His problem, they agreed, was that he tried to play too fast, as fast as Raymond Felton. He was trying to drive a perfectly serviceable Pontiac at Porsche speeds, and QT blew a lot of engines that way. Or, more to the point, he tended to dribble at top speed into the middle of traffic and then lose the ball off an opponent's chest. To reach the same destinations as Felton, Thomas needed to travel at three-quarters of the speed. He had to stop playing as fast as the game seemed, and slow the contest to his natural pace.

That's why it was a little surprising to see him on the floor at a tight moment in such a crucial game, even if it was the first half. Fortunately, Thomas did no harm. If Williams meant the lineup as shock therapy for Felton, who had been spraying the ball erratically around the court in a performance reminiscent of his play in the first Duke game, it didn't work. Felton was in the process of committing four turnovers, and the Heels continued to flounder. At the half, the Spartans led 38 to 33. In the opening period, the Tar Heels had shot one free throw and 12 threes. The offense was out of joint. The next day, the local press fell in love with the word "lollygagging" in describing North Carolina's first half performance. It was an apt word.

With 12 points and two threes, only Jawad Williams had kept the margin from being worse for the Tar Heels. Sean May had shot only two for eight from the field, and picked up a mere three rebounds, easily his worst half of basketball in months. In the cavernous press room, Bob Ryan of *The Boston Globe* and ESPN's Doug Gottlieb, attired in jeans and cowboy boots, were loudly questioning Roy Williams's substitution pattern. Hey, it was one thing for me to wonder about it—I actually cared about the result—but it really ground my nerves when these *interlopers*, these professional *observers*, started doing it. Who did they think they were?

Watching Ryan and Gottlieb and the dozens of other reporters staking out the buffet tables in the media room, I was beginning to think it an accursed thing to be a sportswriter.

You start writing about sports because you love them. But in writing about them, you turn a joy into a profession. You no longer love sports in quite the same way you once did; games are just what you write about. They're your job. And you see everything too closely to be open-hearted about the magic anymore. Cynicism creeps in.

The sportswriters I was meeting no longer enjoyed the pleasures of rooting, the happiness of living and dying through a team. They were partisans only about their jobs. They didn't much care anymore who won or lost, except to the extent that the right team winning might take them to St. Louis in April, instead of to the NIT in New York. Or to nowhere at all. (Good for the golf game, that last option.) The beat writers might like individual players and coaches, but not individual teams.

But if you didn't care who won or lost, why would you want to be around sports at all?

I had come to believe that rather than enshrine the traditional American journalistic value of objectivity, sports sections should hire only homers like me to write about games. Just as once upon a time, the New York *Daily News* delivered the news with a liberal slant, and the *New York Post* fed the appetites of conservatives and everyone knew this, so you might have the Greensboro *News & Record* as the Duke newspaper, while the Raleigh *News & Observer* would be biased in favor of North Carolina. And as for North Carolina State—hmmm . . . did Angier have a daily paper?

I suspected that the subsequent coverage would be harder in all directions; no one dissects a team as ferociously as its own partisan. Just look at the way Crazy Towel Guy talked about Shavlik Randolph. If there are any editors out there who agree with me that this should be the future of sports journalism, give me a call. And let's get this thing started. Down with objectivity! Up with bias! Up with the joys of hatred and the sweetness of love, two passions tenderly intertwined, no matter what they might tell you!

The second half demonstrated once again that North Carolina was a team that functioned by its own mysterious laws and schedules. To the unending frustration of Roy Williams, these Tar Heels often lived less by blue-collar virtues like showing up for work on time than by the aristocrat's assumption of inherent superiority, which allowed them to stroll around in their pajamas (as it were) until past noon. Capable of teasing adversaries with desultory stretches of basketball, as they had Michigan State in the first 20 minutes, North Carolina was also capable of turning around and humbling the opposition with dazzling offensive runs. And that is what the Tar Heels did to Michigan State in the second half, triggered by a Sean May basket that started a 13-to-4 run. Twice the Spartans tied the score but then UNC reeled off 12 consecutive points in less than three minutes for a 61-to-49 lead. Midway through the half, it was essentially over but the shooting. Felton was on his way to finishing with 16 points, May with 22.

Even Melvin got into the act, nailing four free throws late while Bridget screamed from the stands: "Boy, you better hit those free throws for your mama!" In the last month or so of the season, Melvin had tabled his dreams of individual success, or at least redefined them, and appeared to find satisfaction now in being part of something bigger than himself.

With 40 seconds to go, Rashad McCants, who had finished with a quiet 17 points, raised his hands in the air and began pulling his jersey from his chest for the crowd to see. At the end of its second-half rampage, North Carolina had beaten the Spartans 87 to 71.

In the Tar Heel locker room, the players were relaxed, unimpressed by their easeful victory. As the reserve guard Jesse Holley headed into the shower, his towel slipped. "Oh, no, I was almost naked!" he yelled to no one in particular. Melvin was videotaping everybody videotaping him. Reporters pinned Roy Williams against a row of lockers, where he was repeating his mantra for this year's tournament: "Win or lose, the sun will still come up on Tuesday. Win or lose, I'll still be on the first tee that morning."

I don't think he believed it in the slightest.

Back at the Adam's Mark, the hotel where the Tar Heels were staying, Oscar Robertson was sitting at a little table in the bar, smoking a cigar. A Carolina fan took his little girl to meet the legend. "This is the greatest basketball player of all time," he told his daughter.

"That's something, coming from a Carolina fan," Robertson said.

NCAA CHAMPIONSHIP GAME
UNC versus Illinois
Edward Jones Dome
St. Louis, Missouri
April 3–4, 2005
7:40 P.M.

Sunday in St. Louis, the day before the championship, the newspapers in the racks were all from Friday: VIGIL FOR POPE, they proclaimed. But the Pope was dead and so were the downtown streets—old haberdashers, coffee shops, ancient bridal boutiques, buildings with fading logos from the last century: *Lithographing. Stock and Feed.* This was once a cattle town, a jumping-off point for the American beyond. Fur trappers, riverboat gamblers, cattlemen, all ghosts now. Like everywhere else, this was now a convention city.

The Mississippi flowed slowly by, unruffled as a Midwesterner.

At a stoplight idled a station wagon with Missouri plates, bearing a bumper sticker that read: "Hate is not a family value."

That depended on whose family you were talking about.

So the next day, the journalist with limited access rode the press elevator up into the rafters of the Edward Jones Dome, clambered out, and commandeered a seat on the precipice there. Below him were more than 47,000 fans, most of them orange-clad fans of the Illini. The journalist's mood was as sour as the Midwesterners around him were jovial. The elevator operator, for instance, who hadn't seen a speck of sun all day, couldn't have been more radiant with civic decency. "Nice to have you with us," he said. Somewhere nearby must have been the birthplace of the expression: "Have a nice day," which people around here kept saying without a speck of irony.

It had been a long season, for him the equivalent of two in one, given that he'd conscientiously watched all the North Carolina games and all the Duke games, too. So much basketball had almost—almost—exhausted his obsession. He was reminded of a tactic his girlfriend had used years back to curtail her nine-year-old son's burgeoning passion for profanity.

Smartly, she hadn't forbidden cursing and thereby enlarged it to glamorous outlaw status. Instead, she had periodically (usually Friday nights) allowed her son 60 seconds of unbridled swearing. He was free to say whatever obscenities swam into his ken. The boy usually found this freedom so dumbfounding that he locked onto only one or two words, unable in his uncensored minute to say much more than "motherfuckin' shit" (uncontrollable laughter), "motherfuckin' shit" (more laughter), again and again for 60 seconds of foulmouthed joy. And then, for a week or so, he was cured.

More than 70 games in the course of one season had nearly done the same for the journalist in regard to his passion for basketball. *Motherfuckin' shit.* Every game seemed to involve the same thing, guys running up and down court trying to put a leather ball into a basket. Was there anything more to it than that? What if the games were only games, not the portals through which wild, exuberant energies entered and mingled, not the hubs of a thousand intersecting narratives?

The sportswriter from downstate sat beside me. A charming, sandy-haired fellow. "The Rev!" I said, slapping his hand. It was just a form of greeting, like saying "to your health." Down on the court, the Rev himself was tapping fists with the Carolina players right before tip-off.

"The Rev!" he said.

You know white guys are happy when they start to sound like black guys.

Brotherhood was so much more fun than winning. So much more civilized. So much more adult. It seemed that against all odds an adult is what I had become, my values in the right place. I had finally started to put aside childish things. Don't cry for me, Carolina. I made it nearly to 50 before this happened. Now I would step into the ambiguous, gray-skied world of grown-ups, with its civilities and rituals, its sense of proportion. Grown men ought not to be fixated on the doings of their favorite team. Their favorite team ought to be people near at hand, at least no further away than their son's or daughter's Little League team. It took me a lifetime but—adulthood at last! You don't need to drag me there. I'll walk.

And that is how I felt right up to the moment that the game began.

And then, WHAM! Out of the chair I rocketed, screaming. Enough of that sadass shit.

The beast was in the house.

The beast was not going to be satisfied with detachment and maturity. He was going to make sure that the journalist's version of adulthood included some raw, unprocessed emotion.

The vehement passions were so much fun to exercise, probably because they exercised me. I didn't have to do anything. All I had to do was burn and focus like a blowtorch.

For a while, the first half was a game of runs. Illinois drew first blood, then the Tar Heels surged to a 9-to-2 lead, then the Illini scored the next ten straight points. Then UNC went on a 7-to-0 run to take charge once again. From that point, every time Illinois started to edge a little closer, the Tar Heels would race to another commanding advantage. They were aided by the Illini slinging threes at the basket as if they were throwing paint at a wall, hoping something stuck. Not much did; they made only five of 19 attempts from beyond the arc that half, on their way to shooting a season-low 27 percent from the field.

In contrast, Rashad McCants was playing a beautifully economical game, mixing drives and long-range shots, from his first three, which gave Carolina a 5-to-2 lead, to the one that closed the half with 45 seconds to go.

Yesterday, reporters had asked McCants about Luther Head defending him. He had responded, "I don't feel any defender out there is going to bother me one on one." This prompted Roy Williams to jump in and say: "Defense is a team game. Please don't ask my players about Illinois. They haven't even had a chance to go over the scouting report yet." Nice try, Roy, but as soon as McCants went into the so-called breakout room to be interviewed individually, he reiterated: "When I come to play, it's not that pretty on a player that's guarding me."

I had wondered if he might be putting too much pressure on himself, but clearly not. Although he hardly captivated the press like Muhammad Ali, McCants had a little of the boxer in his persona. He psyched himself up by lobbing a grenade out there in the form of a highly charged remark, then diving on it himself. You might say, Why throw the grenade in the first place? Wouldn't it have been easier to keep your trap shut or at least hew to the standard pregame pieties, the way Raymond Felton and the rest of the team did?

But McCants needed doubters, it seemed, so that he could have the vengeful pleasure of showing them up, of rebuking them for their lack of faith. And when he nailed that last three, it looked as if he was going to rebuke his doubters all the way to the Most Outstanding Player award. He already had 14 points and was shooting 50 percent from the field.

I decided that with a 40-to-27 halftime lead against the number one team in America, I had better stay in my seat. I was thirsty and I wouldn't have minded visiting the media room, but what if I did and North Carolina lost? I don't think I could bear the guilt.

"You ain't going nowhere," the beast said.

"Stop telling me what to do," the journalist responded. "I had no intentions of going anywhere."

"I had no intentions," the beast said, mocking me.

If you think that in this modern age, I was as captive to superstition as an Aztec sacrificing prisoners to the sun god so that the light kept returning to the earth every morning, I understand your point of view. But I saw no need to take chances.

I might as well have taken a nip at the break for all the good I had done my team. Though North Carolina ran its lead out as far as 15 points, at 44 to 29, it was too early; only a minute and a half had gone by in the second half, and as the sportswriters say, Illinois was too good a team not to come back.

The Illini began to sink the shots they'd missed in the first half. Deron Williams, Luther Head, and Dee Brown were all playing well. They moved the ball from one side of the court to the other with skip passes, shifting the defense with hardly a dribble. With 12 minutes to play, they had cut the UNC lead to a single basket at 52 to 50.

After Roger Powell hit a layup, the sandy-haired sportswriter from downstate jumped up, held out his hand toward me, and sweetly exclaimed: "The Rev!"

Heavily under the influence of the beast at this point, I looked him off. He pulled his hand back in slow confusion. "Sit down," I told him. "And screw the Rev."

His face flushed and he sat down, appearing not to know what he had done. Oh, I was remorseless in such moments, a terrible beast indeed.

Were it not for Sean May, who appeared to be on his way to fouling out every big man in the state of Illinois, including ones not yet born, the Tar Heels would have been in trouble. McCants had not scored a single point in the second half. May, however, kept muscling shots into the basket from close range, eventually finishing with 26 points and ten rebounds. And yet, for every inside score, Illinois would eventually hit one among a barrage of threes. With only 2:40 to go, Luther Head nailed a three to tie the score at 70 apiece. The sandy-haired sportswriter pumped his fist and simply said, "Yes!"

It got worse. On the next North Carolina possession, the accomplished freshman Marvin Williams missed a jumper. The Illini had the chance to take the lead. So much for my accommodation with losing. Once again, when push came to shove under the basket, I wanted nothing less than what I always wanted: a win.

Fortunately, Deron Williams missed a three. Unfortunately, Rashad McCants missed a daring contortion of a layup. Fortunately, miraculously, wonderfully, Marvin Williams tipped in the errant shot and the Tar Heels assumed a fragile two-point lead. 1:27 was left to play.

Now it was time to stop Illinois. When it came to defense, North Carolina had been erratic throughout the season. Only a week before, Roy

Williams had been so disgusted with his team's defense at the Syracuse regional that he ordered the rims removed from the baskets for the team's practice. On Sunday, the Tar Heels had allowed the normally stolid Wisconsin Badgers to run with them and score 82 points. Wisconsin! A team with maybe one certifiable athlete. Williams had done something similar after the stunning loss to Santa Clara back in November (even now, this defeat boggled the mind), sentencing his team for hours on end to step and slide across the court, locked into their defensive crouch. The implicit message: Shooting is the reward for good defense. And to put it in the terms of a dictum: No defense, no shot.

In a way, Carolina's brilliance on offense often made the humble rigors of defense seem superfluous. Like asking a mathematical genius to balance the family checkbook. Defense was sweeping the kitchen floor, emptying the trash, binding the newspapers for recycling. If the household was to run efficiently, somebody had to do it—hey, how about you?

As recently as two days ago, during the Michigan State game, Roy Williams had felt it necessary to scorch Melvin and Rashad McCants for not picking up Michigan State's Maurice Ager, who had sunk an open three from the corner. "They were pointing at each other like, 'It's not my man,'" the coach said. "I said, 'I didn't care. It's North Carolina's man.'"

For the next minute, North Carolina played defense. The game hung in the balance. The crowd noise rose around us like fire, increasing in pitch as the shot clock wore down. Luther Head missed a three, but the Illini recovered the rebound. They called a time-out, set up a play in which Deron Williams missed a three. This time, the Rev snared the board. "The Rev!" said the sandy-haired sportswriter. He was messing with me. I had just about had enough of the Rev and the sportswriter. Illinois called a 30-second time-out.

On press row, I gnashed my teeth. What a fool I had been to think I had grown beyond caring. The ball was inbounded. Luther Head threw a perimeter pass. And then it happened. Raymond Felton, whose defense could range from brilliant to lax within a single possession, rose towards the ball, nicked it with his forearm, raced after it, corralled it, and headed into the front court, where he was fouled. How many times in NCAA championship history had North Carolina been the beneficiary of such moments? In 1982, Georgetown's Fred Brown mistakenly passed to James Worthy at the end of that game. In 1993, Chris Webber called for a time-out his team didn't have (after a travel that wasn't called), giving the ball back to North Carolina on a technical foul. And now this. Felton hit only one of the free throws, giving Illinois a last chance to tie. Again, Luther Head missed from three, as the Illini had missed the previous nine long-range attempts. Again, Felton was fouled. This time he hit both shots. And North Carolina won the game, 75 to 70.

Streamers fell, strobes flashed, the players fell on top of one another in a prayerful, exultant pile. I headed down underground and came upon a slow majestic procession headed to the locker room to meet the players. Flanked by four Missouri state troopers was Michael Jordan, in snazzy white hat and Carolina-blue sweat ensemble, and Dean Smith, in a business suit, his hair a gleaming silver. I fell in behind them. We passed Jay-Z, on his way to comfort the Illinois players.

The procession and the players arrived at the locker-room doors at exactly the same moment. Smith hugged Jawad Williams, then Raymond Felton. "I'm so proud of you," Smith said, beaming. Melvin was shouting "'82, '93, '05!" Jordan embraced McCants, whispering something in his ear. In the locker room, Melvin videotaped Jordan's and Smith's state visit.

Back in the Dome, "One Shining Moment," the traditional CBS montage celebrating the NCAA tournament just past was playing on the video boards. A picture of Duke's Daniel Ewing crying after the Devils' loss to Michigan State was shown. The Carolina fans took time out from their jubilation to boo. A couple of hours later, the UNC team bus pulled up at the Adam's Mark Hotel. The players poured out to be met by cheering fans, one of whom was holding up a homemade sign asking, "Where's Dook?"

Melvin disembarked to raucous cheers, as did everyone on the bus (had the bus driver also dismounted, he would have been cheered as if he'd been a solid contributor off the bench), found his mother in the crowd, and removed the net from his own neck and as if anointing Bridget with a crown of strings, lowered the net over her head. "I love you, Mom," he told her. And he picked his way through the roar of fans and disappeared upstairs to the players' floor.

Bridget was besieged by Carolina fans. Everyone wanted to touch the net around her neck as if it were a piece of the True Cross. "Whose mother are you?" asked an older woman with frothy hair dyed Tar Heel blue. She put her arms around Bridget and kissed her.

Bridget ducked behind a police barricade to smoke a cigarette, her only vice, she said, besides playing the lottery. A Carolina student approached her. "Are you Melvin Scott's mom?" he asked her. Proudly, she shook her head yes, blowing out a long stream of smoke.

"He may not end up an NBA star," the student said, "but he's a good guy. And he's my year!"

Bridget bridled. "He may too make the NBA!" she said, her voice rising. "You don't know what he's gonna do."

The student's eyes widened. He backed away, his arms up in the air, signaling no contest.

She shook her head in disgust. "He doesn't know my son's capabilities."

She strolled back into the hotel, headed upstairs. "Melvin showed the negativity in Baltimore," Bridget crowed. "Yes, he did. He answered all the negativity!"

By the elevator, a guy stopped her and offered to buy the net from her. "I'm not selling this," she said. "You're crazy! This is going back to Baltimore!"

Upstairs in Melvin's room, a group of friends and family were hanging out, waiting for Melvin to return. Everybody was watching ESPN. A feature on Roy Williams aired. The room went silent. Then Melvin arrived and everybody broke into applause. "You played good defense on Brown, son," Bridget yelled from across the room, referring to Melvin's effort against Dee Brown.

"That Brown boy is fast," Melvin said. "That nigger scared the hell out of me. Excuse me, Lord."

He joked about "the Reverend Powell." "The *Reverend*. The Reverend be throwing 'bows,'" Melvin said, referring to Roger Powell's muscular sermonizing inside the lane.

He handed out juice to everyone in the room, then stood there, watching ESPN while everyone else watched him. A shot came on of thousands of students back in Chapel Hill, celebrating the victory on Franklin Street. Young men were vaulting over enormous bonfires, as if toying with the idea of sacrificing themselves in gratitude to the fickle fire gods of basketball. "Y'all think this is a joke," Melvin said. "But they jumpin' through fire!"

Ebulliently free-associating, Melvin announced: "Michael Jordan. That nigger hugged me." He handed his mother his miniature video camera so that she could watch Jordan's and Dean Smith's visit to the locker room.

Digger Phelps appeared on the television screen. Everyone in the room booed. "Digger Phelps and all those guys were against us," Melvin said. They had turned this to psychological advantage, playing with a chip on their shoulders, which is usually a hard thing to do when you're the number two team in the country. But the commentators had heaped fuel on a simmering resentment. Their view was that Carolina was a collection of talent, not a team. Add to that perception Phelps's gall in predicting an Illinois victory, a prediction the players had seen replayed endlessly as they lolled in their rooms, waiting for the last game.

"That three you tried to take, that was a good shot," Bridget said.

One of the Baltimoreans in the room said gruffly, "I told you how to shoot that!"

In a stagy aside, Melvin said, "At least, I don't have to hear that anymore." Then he said proudly: "They gave me a lot of credit for keeping the team going when Felton was out."

"We live in Hate Town, USA," said the same Baltimore guy. "But the critics are not calling now. My phone used to be ringing off the hook, asking about Melvin. But not now. This memory is forever."

It was hard to tell how Melvin reacted to this somewhat equivocal assertion of triumph. He disappeared into an adjoining room, stripping off his shirt and tie, and then returned, clad in his wife-beater. He combed through his dresser, looking for the right outfit. He disappeared again into the other room and returned wearing his going-out duds: gray shirt and pants and the pièce de résistance, the blinking belt, which in bright green letters kept spelling out "NATIONAL CHAMPION."

Digger Phelps reappeared on the TV screen. "Digger's sick," Melvin said. "Look at him. He's hurting now. He was killing us. Get out of here, Digger Phelps!"

Jawad Williams came into the room, pointed at Melvin, and said, "This is my son." As the two players exited into the night to celebrate on the town, Bridget said, "I know y'all national champions, but don't y'all get too tipsy."

By the side of the hotel waited a stretch Hummer, rented for the occasion by former UNC football stars Julius Peppers (who also played basketball) and Dré Bly, both now professionals in the NFL. To the raucous cheers of drunken UNC fans, Melvin and Jawad disappeared into the Hummer. About to head off herself to a cheaper hotel out by the airport, Bridget watched them go. "They don't know what that boy can do," she proclaimed to no one in particular. "He came from 20th and Barclay."

I wished everyone a good night and went back to the hotel. There was one more thing I had to do. I sat on the floor next to the bathroom in the muted light of the TV, showing its endless clips of Sean May pumping his fist into the air at the moment of triumph, and I called my mother.

It was two in the morning. She would have been lying on the couch.

She answered the phone on first ring. "Whooohooo!" she said.

"Wasn't that amazing?" I asked.

"Oh, it was simply wonderful!" she said.

"Did you go to Franklin Street?" In 1982, she and I and my ex-wife had walked uptown after Dean Smith's first championship. A naked student painted in blue had dropped from a tree and planted a big kiss on my mother. "What are you doing out at this hour?" he'd asked her. That was 23 years ago.

"No," she said. "Not this time."

"Well, it was amazing," I said.

"Just wonderful," she said.

We talked on into the night, about the game, about the season, about how wonderfully it had all turned out.

THIRTEEN

In the Sun of Their Time

GRADUATION

"I'M ALREADY CRYING," Bridget said. She wore a suit, and her fingernails were painted white. It was 9:27 on May 15, three minutes before the start of graduation. Melvin's family and friends had assembled at Kenan Stadium for the outdoor graduation ceremony. It was a muggy May morning, classic Chapel Hill commencement weather. Tiny red bugs were crawling over all of the seats. Resplendent in striped shirt and latte-colored pants, Charles didn't want to sit down on the insects. "Yo, y'all, I'll sit like this," he announced, perching on the seatback. His hair was as intricately braided as a princess at the court of the Sun King.

Bridget kept saying "20th and Barclay, 20th and Barclay," a reminder of the distance Melvin—and by extension, the rest of the family—had come to be here today.

Here on their fiftieth reunion, members of the class of 1955 straggled by like mountaineers roped to each other on a steep descent. They waved at the crowd. Bridget waved enthusiastically back at them. Her joy was flowing in all directions. Just then, her phone rang. It was Melvin. "Hi, Sweetie!" she said brightly. Melvin wanted to know if everyone had arrived.

The graduating students began to stream down the stairways into the end zone seats at one end of the stadium. Before anyone else, Bridget jumped to her feet, applauding. Wearing her thick black glasses, she scanned the procession for her son. Now Melvin was on the phone to Charles. "No, look to the left," Charles commanded. "We're right here." Finally, with mutual location visually confirmed, Charles raised his fist in the air, and across the way, everyone watched Melvin jumping up and down in his blue gown and occasionally mixing in a little Baltimore dance move.

"Before you sit down, man," Charles warned Melvin. "Watch your seat. There's a bunch of red little things."

Now Melvin appeared to be conducting the orchestra. He blew a kiss across the stadium, then put his arms in the air and waved them back and forth. He stood out even in a crowd of nearly 3,000 graduates. Back on the cell phone, he asked his mother, "Ma, why do you keep waving? I see you already." Before hanging up, he said to her: "All right, baby, I love you." Bridget beamed.

She applauded for everything, the march of the faculty, the conferral of honorary degrees, even the beach balls that went bouncing across the heads of the seniors. She clapped when Nan Keohane, the recently departed president of Duke, received an honorary degree. By contrast, the senior class welcomed her with a thunderous boo. The rivalry was never far away, even in such genteel moments.

The speaker for the occasion was Peter Gomes of the Memorial Church at Harvard. His first piece of advice to the graduating students was "Toss your balls while you can." And then he told them, "Cherish your failures. Most of us learn more from our failures. You will not always win, you will not always be first. Failure is inevitable in life, so learn from it. Don't bet on celebrity. Invest in character. It lasts a lot longer."

Bridget had been listening intently. "The Scotts are not losers," she said. "We are fighters. We may not win, but we'll fight."

With that, it was time for the conferral of degrees en masse. Melvin stood up and once again appeared to be conducting the students as cries of "TAR" rose from one half of the graduates, to be answered by the other half with a resounding "HEELS." He was now a graduate of the University of North Carolina, with a degree in African and Afro-American studies. The class president and vice president performed a scrawny if good-humored rendition of "Don't Even Worry, We're Going to Make It," the song the team had composed and sung together during the NCAAs.

At the end of the ceremony, the crowd sang the school song, "Hark the Sound," which was followed, as it so often is, by a completely spontaneous roar of "GO TO HELL, DUKE!"

RUBY TUESDAY'S

After the commencement, everybody piled into cars and drove to Ruby Tuesday's on Highway 54. Fellow graduate Jawad Williams showed up with his family, who'd arrived from Cleveland. The hostess asked Melvin if he had a reservation for everyone. "Yeah," Melvin said. "Dean Smith called it in." The hostess scrutinized her reservations book and couldn't find any mention of Smith or Scott. "You seat us," Melvin said, smiling and changing tack, "and you're gonna get the biggest tip in history. Your pocket'll be so big it'll be like it got the mumps."

On the wall behind the hostess hung a Duke jersey. Nobody paid it any mind. While the Scotts waited for a table, Charles went to the salad bar and helped himself.

Now that Melvin had graduated, his scholarship was over. And his Pell grant was running out at the end of the month. "That's all right. I'm a survivor," Melvin said. Maybe better than being a survivor was the good fortune that his new landlord happened to be a Carolina fan who had greatly admired Melvin's performance against Villanova in the regionals. So much so, in fact, that he had decided to let Melvin live rent-free for the next 12 months in an apartment he owned in Chapel Hill. By coincidence, the landlord and his wife and daughter were leaving Ruby Tuesday's while the Scotts were waiting for a table.

The man asked Melvin how it was going with the apartment.

"The air's not getting up to the second floor," Melvin said.

"Maybe you're leaving the front door open when you're moving all your stuff in," the landlord said.

"Nah," said Melvin.

"I'll see what I can do, Melvin," the man said. "Congratulations on your graduation."

By the time the families were seated, Charles had eaten so much salad that he was nearly full. Jawad's family and friends took up one half of a large table in the middle of the dining room, the Scott clan the other. The Williamses were mostly quiet, the Scotts mostly raucous, having a terrific time laughing and telling stories on each other. The boys— Melvin, Charles, and Will (Kevin was in boot camp in Maryland)—were trying to top each other with a tried-and-true repertory of tall tales, sports arguments, and past wildnesses. Bridget was the delighted audience, surveying her family like the Queen of Baltimore.

"Uncle Melvin," Tay asked from across the table. "Who was tougher in football, you or Uncle Charles?"

Charles snorted. Melvin looked across at Tay, shielding his mouth from Charles. *Charles is crazy*, he mouthed, circling his ear with his finger. *Crazy.*

Charles's response was to begin talking smack about how he would have put a hurting on the North Carolina wide receiver Jesse Holley, who also served as a reserve on the basketball team. "He might bust me at the line," Charles said, "but I'd bust him right back."

With that, Charles and Melvin started pantomiming the moves they'd make at the line of scrimmage, the feints and uppercuts they'd deploy on their opponents. Then it was on to recollecting fine old times. Charles said: "I remember one time I wanted a bike and didn't have no money." The Scott half of the table broke into whoops and hollers with Charles's intro. They clearly knew what was coming.

"*No money*," Charles emphasized. "So I stole a bike, stopped to go in the store and buy some candy, and when I got out, someone stole the bike from *me*!" Everyone roared with laughter. Charles look bemused at the way justice was sometimes dispensed in the universe, Baltimore-style. "So I decided on another plan."

"Charles and his little gang," Bridget said, smiling.

It was a perfect plan, Charles said. They entered a department store out in Towson, a nearby community, and hid themselves in clothes racks so that they would still be in the store after it closed for the night. So far, so good. The employees went home and the boys emerged to pick out the bikes of their dreams, gleaming, new bikes, and wheel them out into the liberty of a Maryland night. Only one problem. They hadn't figured on the night watchman.

"I told him we were there to get cigarettes for our family," Charles said.

Back in the city, Bridget received a call at three in the morning from the Towson police. "We have your son," they told her. "If you want him, you're going to have to come get him."

"Those were some days," Charles said.

Melvin told the story of Moogie. Once he and his buddies were walking in Baltimore. They spotted a fellow wearing a homemade white jersey that said: *Melvin Scott. North Carolina. #1.*

"I figured the least I could do was show him some love," Melvin said. He walked up to the man. "I told him, 'Yo, that's me. Thank you.'"

The man looked at Melvin. "That's not you," he said. "That's Moogie."

Charles and Will, who was wearing a black Samos shirt that read "We On Fire," fell out laughing as Melvin told the story.

"No," Melvin insisted. "That's me!"

"That's not you," the man repeated. "That's Moogie! I know him!"

At the table in Ruby Tuesday's, Melvin laughed. "I gave up," he said. "I said, 'All right, man. That's Moogie, whoever Moogie is.'"

BLOOD IS THICKER

Lunch ended in the middle of the afternoon, and the brothers decided to play pickup at the Dean Dome. Tay tagged along with his uncles. At the practice gym, the games were hard-core, lots of collisions and cursing. The Scott boys were joined by Jesse Holley, the football star and basketball reserve, as well as another friend of theirs from Baltimore, known as Wookie. Melvin went out in the first game with an injured elbow, which he now held tenderly on the sidelines as he watched the action.

"Don't cheat on my bros," he yelled at Jesse Holley, who was involved in a scorekeeping dispute with Charles. "Blood is thicker."

Holley scored against Charles on a drive. "Fuck!" shrieked Charles.

Holley came over to Melvin, leaned down, and nodding his head towards Charles, said, "He's a Baltimore rottweiler, isn't he?"

The game was getting rougher. Holley and Wookie were playing Will and Charles. Will Scott moved with broadband speed, and Charles was playing with his usual ferocity. All the Scotts could hoop. Smitty had described Will as conspicuously talented, but lacking Melvin's work ethic.

During the next break in the action, it was Wookie who drifted over to Melvin. He held up his arm, which was slashed bloody. "You'd think I was a dope addict," he said. "All these scratches on me."

Charles drove hard to the basket, got banged around, and scored a layup. He came down, clenched his fists, and howled. On the next play, he hovered outside. The defense converged on Will barreling to the hoop and he fired a pass to Charles who knocked down a three. "That was the difference in our high school careers," Melvin called to Charles. "You was a spot-up shooter for Smitty. I was a jump shooter in people's faces."

Melvin could be lionhearted, too. Charles just grinned. He knew it was true.

At the other end of the court, under Melvin's watchful supervision, Tay was playing one on one and shooting. "Go down there and shoot runners," Melvin told him. Tay did as his uncle asked. The boy liked to get the ball in triple-threat position and bait the opposition by sneering, "Say hello to my *leetle* friend." Then he'd either drive or launch a three-pointer, which he slung from his hip but which frequently went in. Tay's crossover was especially nasty; he swung the ball from one side to the other as if he were mesmerizing it.

"He's a better point guard at his age than I was," Melvin said. The word from Baltimore was that Tay passed with more skill than the young Sebastian Telfair, but that his big men—however big they got in the sixth grade—couldn't finish. They stayed atop of such information in Baltimore.

At just that moment, Jesse Holley came streaking in to try and block Tay's three-pointer from the corner. Tay pump-faked him and let the shot fly. Swish. Melvin smiled. The next generation was on its way. The cycle was beginning again.

YOU'LL PUT ME IN THERE ONE DAY

One fall, my father and I were driving to the farm his mother had grown up on in Virginia. "At your age," my father told me, "I had a job, a house, four kids, and a wife."

"I remember," I said. "I was one of the seven."

A barren, involuntary smile appeared for an instant. "Why did I have to raise such rebellious children?" he said. "Not one of you goes to church, except for John."

"I guess that's because we take after you," I said. I meant this as a tribute, and normally he might have winced with crinkly-eyed pleasure at such a remark, but not today. "No, you didn't take after me," he said dolefully. He looked disconsolate, consumed by thoughts he didn't wish to share.

Deep in the country, we passed by a tiny family cemetery, shaded by a single oak, and guarded by a rusting, falling-down fence. The headstones rose out of the ground at odd, crooked angles like a bottom row of bad teeth. A November rain glazed the graves with a lonely sheen.

"One of these days, you'll put me in there," my father told me. He could be theatrical in his despairs, so much so that I sometimes felt I needed to inject a note of levity into his complaint.

"Not that particular graveyard," I said. "That would be whatever the opposite of grave robbing is."

He didn't laugh. Truth be told, there were times I couldn't wait to bury him. I didn't know whether this was the condition of all sons, or only the particular grievance of this one. In death, it seemed to me, I would be able to wreathe my recollection of him in the sweet smoke of nostalgia (the burning leaves of October). In life, he could be impossible. But in many ways, his mortality seemed a moot point, as he didn't appear like a man capable of actually dying, as unscientific as that intuition might have been.

Where would all that force—his great and difficult gift to us—end up?

After a little more silence had passed, I said, "Not any time soon, I'm sure."

"But just the same, you'll have to one day," he said. And now I know that this is what sons do for their fathers if the fathers train them well. They bury them.

DYING TO REMEMBER

"Remembering everything is a form of madness."
—*Translations*, BRIAN FRIEL

This compulsive remembering . . . I remember my father remembering . . . at night, in the living room, after supper . . . the pipe bowl afire like a volcano seen from the air. . . . He was trying to tell me something with all these stories . . . trying to bring up these glittering wet fragments of time like a fisherman hauling up a net of fish . . . but it wasn't the stories themselves—he was trying to teach me the necessity of recollecting.

And all of this remembering hadn't been enough to save him. He was, astonishingly, dead. Dead. No word for that, really. No story could sufficiently convey that. He was beyond his own remembering but not beyond our own and perhaps that was his intent. To stay with us in the nostalgic enclave of our hearts. But dead. If he was in the *bardo*, as the Tibetan Buddhists believed, he had 40 days of confusion to go before he became a cardinal.

In the hospital, I turned back to look at him one last time (the motion of memory) and it wasn't him. He had left a sphinx on top of a hospital bed. I didn't recognize his profile. My brother, David, and I signed the form releasing his body to the funeral home. We went home and had a stiff drink with the rest of the family, raised to his soul, wherever it might be, and I slept on the floor next to the couch where my mother fitfully slept, unwilling to go back to the bed they'd shared.

I remembered all of this for all of the good it had done me.

HOME

At last, the season was over. For the first time in months, there were no games to follow, and I found myself slowly returning to human form.

The beast in me was hibernating, satiated beyond his wildest dreams by that Monday night in St. Louis. For a brief moment, he had considered what it might be like to feel solidarity with, even compassion for, his Duke antagonists. The moment passed. As it was, the summer was tradionally a slow season for the beast. Baseball bored him. Especially baseball writing, those elegies to green fields, our nation's rural past, all those spike-shoed country boys like Ty Cobb running around the base paths with blood in their eyes. So, at present, the beast slumbered, never uttering a discouraging word. With the resumption of basketball practice in October, he would be awake and ravenous.

Without games to cover or the beast to feed, the journalist found himself at loose ends, straining to understand the season he'd just been through. It had raced past in a whirl of sensation. He had gone home again after so many years in New York. Memories kept erupting like gushers in a West Texas oil patch; from the grave, his father kept right on talking, telling him what to say, what to think, how much he loved this place, how fragile were its best values. In a way, the season had been a dialogue with a dead man. The journalist knew what he would say in every situation.

On the one hand, the games he'd watched had been merely games, had they not? And it was foolish for a grown man to care that much about mere games. Or, so the journalist tried to tell himself. On the other hand, any other phenomenon that so stirred a fellow would have been scrutinized by sociologists and historians and pundits of various stripes. It wasn't as if he were the only spectator who felt this strongly about a sports rivalry. Basketball marked his life like the rings on a tree, from the soundless broadcast on Channel Four of a UNC game he'd watched sitting between his parents on a snowy night in the Sixties, to the championship bout in St. Louis, which he'd caught from the dirigible-high press box. Why not study his own obsession, make a scholarly foray into a subject that he never failed to find mysterious: himself?

He decided that a season of great basketball and consummately awful poetry (with exceptions) deserved to end with a real poet. And he knew just the man. Felicitously, he happened to be a professor at Duke who'd kept a long, informed watch over the goings-on in this part of the world, including the mania for basketball, which he shared. He had been raised in a North Carolina country town, and he loved the South. And if his poetry was any indication, he'd been ruminating on the role of memory in this place for many years. His name was James Applewhite.

He lived with his wife outside of Durham in a house nestled against a dark benevolence of trees. The house sitting in the woods gave a sense of memory holding out against the omnivorous encroachment of time. Leading me through the light and airy house to the back porch, Applewhite told me that they'd lost many trees when Hurricane Fran blasted through in 1996.

He said that he hardly remembered the hurricane. He and his wife had been up all night talking with their son about his divorce. The hurricane was a loud afterthought. Applewhite put his head in his hands at the memory.

His gout was acting up on this glorious spring afternoon. Just like my father's used to do. A slender man wearing khaki shorts and a sports shirt, he propped his foot up on a stool, and kept replenishing his glass of water. "They say it helps," he said. "It's what they say, anyway." A wry shrug, as if any wholesale proposition should be politely answered by a bit of skepticism, a characteristically Southern reaction.

He'd grown up in the eastern town of Stantonsburg, about ten miles from Wilson, North Carolina, a place that used to call itself "the world's greatest tobacco market." His grandfather moved into town from the family farm, kept a general store. His father had run a gas station and a garage, and later sold electrical appliances. They were both Methodists. Applewhite's grandfather, in particular, venerated the local and the elemental in a way that constituted a religion of the-near-at-hand. "There was this sort of sacramental valuing of white scuppernongs," Applewhite said. "Or an apple. Or water we'd drive to Seven Springs and bring back in these big five-gallon demijohns. The reality of things was imprinted with a clarity that I don't think people in urban environments would get in quite that way. There was very much a sense that all material things were a gift. A gift from God ultimately."

Like my father, the Applewhites revered tradition. "My grandparents practiced what V.S. Naipaul called a 'religion of the past,'" he said. "They inculcated in me a reverence for tradition, for the way things had come to be."

This could have been my father talking, his life. Memory emerged out of love. And it was these very memories, the containers for the missing sacraments of scuppernongs and spring water and country hams that were threatened by the new North Carolina, a place that seemed to be arriving first and foremost at Duke University. Or so my father believed.

Applewhite struck me as a man of noble sadnesses, steeped in the strong tea of his memories. From his youth, he appeared to have had an elegiac cast of mind and perhaps to his astonishment, had now aged into the decades when losses arrived almost on schedule, like a train pulling into a lonely country station at night.

I asked Applewhite where that obsessive desire to remember and commemorate had come from. "You can contrast the national paradigm of mobility—the westward horizon—with the settledness of Southerners, their largely agrarian patterns," he said. "Being raised in small places, small towns, maybe with fairly slender material means, attaches you more deeply to your immediate physical, cultural, and familial surroundings. There was great deprivation after the Civil War, which meant that anything a family did have, including its own status, was very much to be cherished. And remembered."

Applewhite had gone off to Duke in the early Fifties with the idea of being either a scientist or a writer. His father had attended Duke, so he went there, too. "When you're a Southern kid," he said, laughing, "you're faced with all of these givens. And this was one of those givens. You don't know where any of them come from, but that's the way it is." Upon his arrival at the university, he experienced "a kind of culture shock." He was a long way from scuppernongs and spring water. "It seemed like everyone there was from New Jersey," he said. "Or somewhere equivalent."

He had been raised with what he described as the kind of aw-shucks, down-home, no-artifice demeanor that country kids tended to display. The fraternity life made him uneasy. "One part of me felt as good as anybody," Applewhite said. "But in another part of myself, I felt diffident in relation to the kind of Northeastern, big-money social elitism that was prevalent in the Greek system."

Eventually, the poet beat out the scientist for Applewhite's prime loyalty. He had come under the spell of a legendary writing teacher named William Blackburn, who also taught such notables as William Styron, Anne Tyler, Reynolds Price, and Fred Chappell. By example and teaching, he showed Applewhite how writing—a life of words—could be a manly occupation for a Southern boy. After he finished his undergraduate studies, Applewhite went on to informally study verse for a few years with Randall Jarrell at the University of North Carolina at Greensboro. Eventually, armed with postgraduate degrees, he made his way back to Duke in the late Sixties, and began teaching there.

Throughout his years, he had followed the local sports teams with the natural avidity of a North Carolinian. As a boy in the Forties, the decade before basketball made its mark on the region, he remembered having rooted for the North Carolina football team and its star, the All-American tailback Charlie Justice. "The idea of a Southern team having a national star was intoxicating," he said. "It was us against the world." His father went off to one Duke–Carolina contest, leaving Applewhite at home to listen on the radio. It was a hard-fought game and Carolina won. He was happy. His father was not. "He came home and he was really mad," Applewhite said, eyes twinkling. "It didn't help the father-son relationship."

When he was an undergraduate at Duke, basketball there was, to use his expression, "small potatoes." The hatred between fans of the two schools seemed of relatively recent vintage, he thought—since the arrival of Mike Krzyzewski. "Before that," he said, "Duke hadn't been good enough to be the object of hate for very many years."

I found it refreshing to talk with a man who had clearly watched as much ball as he had read Wallace Stevens. He felt that there were great similarities between Duke and North Carolina in the value systems their coaches instilled. "If you put Mike Krzyzewski and Dean Smith side-by-side," he said, "and you think about it: They stress team play; they stress the authority of the coach; they stress loyalty and earning the right to start. And they have created lineages of former players and coaches who are now out there in the world approaching the game the same way."

People these days were looking for verities and inalterable values,

Applewhite said, and this was evident even in the clash between Carolina and Duke, the culture war played out in basketball. "I come from a sufficiently sacramental, old-fashioned culture to know that folks who came into the world believing are very much at sea in this modern world. And they're struggling to find points where they can make a stand."

This struck me as wise, a reason that symbolic issues generated such fire. People who had grown up in the vast lull of a slower age that must have seemed timeless sensed the world changing uneasily around them. In response, my father had looked for a way to stop time. Even to reverse it.

I explained to Applewhite how Duke seemed like a way station for the upwardly, speedily mobile, the antithesis of the rootedness and love of place of which he was speaking.

"There's some truth in that," he said.

I asked him about how he saw Duke's place within the culture of North Carolina. "In *Gulliver's Travels*," he responded, "one of the voyages is to this floating island which is inhabited by these exquisitely intelligent people, whom I took to be like scholars. And Duke is like the floating island that didn't quite touch the soil of North Carolina."

It was a lovely metaphor: Duke aloft over the red clay of its home state, there but not there, a bubble of Northeastern cerebration wafting in the humid air, unwilling or unable to touch down. Applewhite attempted to qualify the notion slightly: "I don't think Duke is comfortable thinking of itself as a North Carolina school. In its own self-evaluation, it hovers in a kind of Ivy League terrain. Kind of like an electron that can be in several places at once." Now Applewhite spoke as a poet inhabited by the physicist manqué.

"Why do you think Duke would prefer to hover as opposed to sinking down roots?" I asked.

"It's a *nouveau riche* university," he answered. "And it's not even that *riche*. It compares itself with Harvard and Princeton. The only thing it's really ever had as self-definition was the aspiration towards academic excellence. And academic excellence in the U.S. has tended to be defined by the Ivies."

Nouveau riche Duke might be, but Applewhite's feelings towards his school were those of a venerable lover. "Let's say I teach a late afternoon seminar," he proposed. "And I come outside. And maybe it's late October, early November. Maybe the leaves are off the trees. I can see the inky lines of the branches of those old oak trees above the library, of which I particularly like the architecture. And maybe I see the Duke Chapel, which I don't particularly like because it's actually an enlarged parish church blown up to dimensions of a cathedral. But leaving that aside . . . I see the top of the tower of the chapel over the roof of the library as I'm walking. And this is the place where William Blackburn taught . . ."

He stopped and turned his head. His voice was choked. He was crying. His evocation of a stroll across the campus late on a fall day had aroused beloved memories. Truly this man could have been my father.

"And it's where I went to school," he said, trying to continue. Again, he stopped. He wiped his eyes.

"These foundations go down into the soil," he said. "That's rooted."

Applewhite's foot was hurting again, so he got up to hobble into the kitchen to pour himself another tall glass of water. He returned, repositioned his foot on the stool as if it belonged to someone else, and slowly sipped his way into the dusk. A fine spring evening approached. Birds flitted through the darkening woods. They sang to us from the trees overhanging the porch. My father would have known their songs. Applewhite and I dawdled on the deck in the North Carolina dusk, not saying anything for a moment, letting the dusk come on.

He decided to tell me one last story about the pull of place in this part of the world. It involved Robert E. Lee, whose head—or, at least a version of it—was at this very instant in a little box under my bed.

"You know how Lee is entombed in the chapel up at Washington and Lee in Virginia?" he asked.

I did.

"And they got his whole family around him," Applewhite said. "Well, he had had a daughter who had escaped to North Carolina and gotten married. When she died, she was buried down here. And the Friends of Confederate Veterans, whatever they're called, managed to get her dug up and brought back to Virginia to be with her family." Applewhite laughed.

"Now that is Southern gravity," he said. "You can't give up your home place, you can't let the children escape." He didn't need to add, You can't let the children escape, whether they—or you—were living or dead.

I knew the feeling. After all, I was home.

> *Voices of a final hymn, choruses completed,*
> *ascend from the particular Sunday, free and*
> *heard only perfectly over the finished village.*
> *Below, the railroad steel shines evenly,*
> *uninterrupting, carrying boxcars of air, past*
> *bodies discovering the lines of aging, a cemetery,*
> *and on into another horizon. As the imagined*
> *train is leaving, the model village*
> *embodies enormous longing. The desire of spirits*
> *is to exist again: to write new names*
> *with house-rectangles, in windows overlooking fields;*
> *to cut these resembling, different silhouettes*
> *into the horizon, like hats on a Sunday morning:*
> *shed of a sawmill, tin planes reflecting*
> *late daylight, water-tower gaunt in the air,*
> *like a people pictured in the sun of their time.*

—JAMES APPLEWHITE, FROM "THE WIDOW"

MY FATHER AS A BIRD

My father loved birds, and in his sleep, he often possessed the gift of flight. He awoke from his dreams to tell my mother about how he had flown above the rooftops of our neighborhood, over the woods, past the plate-glass windows of the contemporary houses set into the hillsides, above the dormitories and the classrooms and the hospital where he worked and would one day die. Husband, father, professor, those were the requirements of his days; his nights were taken up with cavorting in the sky, swooping, diving, soaring, twisting and turning free.

"I wouldn't mind coming back as a bird," he told her on many occasions. "Zip around, *zwoop, zwoop*." He imitated a bird in flight, his lips pursed to make the sound.

He studied birds, owned a shelf overflowing with manuals and guides. He knew their songs. At the supper table on the back porch on summer nights, he offered us money if we could identify the bird singing in the dusk. "Summer tanager," I usually said. I didn't remember the song, but I did remember that the answer to these impromptu quizzes was often summer tanager. I was already deep into basketball. Birds were birds. I liked them, but I didn't need to distinguish among them. Let a thousand birds sing. In this, I was a grievous disappointment to my father.

"You're guessing," he accused me.

"No, I wasn't," I said. He kept the money.

An old man who flew above the rooftops of a city in the night was less likely to be impressed by 18-year-olds with 42-inch verticals. He got up far higher than that. He could stay in the air all night. His hang time was forever.

An Apology to the Reader
(make that an explanation)

HAVING PRODUCED EXACTLY one book in close to half a century, and that being the one you're about to close (if you're not actually reading the afterword first, you dirty sneak), I thought I had better put in everything related to the topic of Duke-Carolina basketball, because given my torpid rate of productivity, I might not get another chance. While three of my four grandparents made it into their nineties, my father died at seventy-two. And I suspect that I take after him in many ways, despite our differences. We share the same complicated blood. So I wedged everything in here, like a traveler running late, stuffing his suitcase willy-nilly. This explanation may come too late to mollify those of you who bounced too hard around the ramshackle precincts of this book.

"Just be sure you don't forget to put some basketball in there," my editor used to say during the writing. "It is supposed to be a book about basketball, after all."

"Oh yeah, that's right," I told him reassuringly. "Don't worry. There will be plenty of basketball. I mean, of course! Man, come on!"

Well, you see what happened. What I wrote is partly memoir, partly history, partly the chronicle of a season, partly a fan's notes, partly profile, partly crackpot theorizing, and partly other people's poetry. The danger, of course, is that a book that tries to be about everything ends up being a book about nothing. But I really couldn't have written this any other way. My mind seems to function in a highly allusive manner. That's the grand way to put it. Another way might be to say that I am easily distracted, as if I had smoked a lot of hashish (but in fact, none was consumed in the preparation of this manuscript). For me, one thing leads to another which leads to another which leads to another which eventually loops back around to the original notion.

If I'm not mistaken, that's the nature of obsession, that constant circling and ruminating. And indeed, my father was a ruminator—as the men in our family tend to be—and I realized early in my life, by the age of five or six, lying under the sofa lost in reverie for the previous year of kindergarten, that I was also a ruminator. The women are another story. For a while, a ruminator in the thrall of his obsessions can stop time, like a man in an opium stupor. Let's say he's a nostalgist like my father. The sweet smoke (October, burning leaves) of the past swirls around him. He inhales at his own risk. The past returns but the rest of the world keeps moving forward, and he might wake up to find himself having slept by the side of the interstate. What's he doing here?

So the question becomes: How do you safely collect that autumn smoke? How do you transport the valuables—the values, to be precise—of your history into the future, so that your children might benefit from them? And so that you might not feel so lonesome as the generations change and the world and maybe even your own children start to look unrecognizable, more strangers than kin? That appears to have been my father's dilemma.

My own difficulties were related—how to separate a father's wisdom from his guff, the genuine from the malarkey. How was I to use the past without getting trapped there? Often, love's tyranny and benevolence are wrapped together in the same package. How do you disentangle them? One astute reviewer of the book suggested that I was having a conversation with my father, trying to show him that at heart, we shared the same essential values.

Out of politeness, which was much valued by my father, I'd like to give him the last word here. He'd be surprised, as he rarely got that with me when he was alive.

The word, by the way, is pecan.

We were in Connecticut, the Blythe family, on our way north to Boston and my mother's people; this was more than forty years ago. My father must have been in his thirties; I was a boy. The more exotic the terrain, the more Southern my father became (I would learn later to do that same old soft-shoe in seersucker and bucks). And Connecticut was plenty exotic to us gaping Carolinians—dimpled Revolutionary War cannons on village greens, lobster shacks, Congregationalist churches. The air smelled of salt water and lawns. The woods were dark.

We stopped for supper in a diner just off the New England Thruway. Diners were exotic, too. We had none of those in North Carolina. For dessert, we all decided on pecan ice cream. My father did the ordering. The young waitress corrected his pronunciation of pecan. "It's pee-can," she said, rhyming the last cringing syllable rather harshly with "man." Child that I was, I detected an alarming sourness in this woman, who was otherwise attractive—maybe she hated having to wrap her hair in a net and wear stockings and a white uniform with green piping. Maybe she hated being a waitress. Maybe she hated us. But if so, hatred seemed not to bring her any happiness. She just wasn't having any of this pecan business.

"I guess you don't know," my father said teasingly, always ready to banter with a pretty woman. "It's actually pronounced puh-khan. But you're too young to know about that." My father's grandfather used to say that a pee-can was something you put under the bed for when nature called in the middle of the night. Clearly this pecan business had aggravated many generations of Blythes.

"Pee-can ice cream," the waitress repeated. "I'll be happy to bring you some pee-can."

"I would prefer it very much if you brought me some puh-khan ice cream, please ma'am." My father's smile tightened. I could hear the furnace of his temper beginning to clank inside of him. He hated rudeness and he regarded the North as an unaccountably rude place. He was always on watch for a Yankee to do something graceless. And now this pretty waitress had done it, had insisted that she was right and he was wrong (what kind of native treated an outlander this way?). The word, she told my father again was "pee-can."

"Puh-khan, please ma'am."

We were behind him one hundred percent in his titanic struggles, even though it was starting to look bad for dessert. I considered bailing out for vanilla, a word that was pronounced equally in Connecticut and North Carolina as far as I knew, but I couldn't bear the disloyalty. If I was growing up to be a traitor within the family, aware of my father's maddening contradictions, I was no traitor outside of it. Not yet, anyway. Not here in the belligerent green lawn state of Connecticut.

"I would be just delighted to bring you some pee-can ice cream," the waitress said.

"Puh-khan, please ma'am."

"Pee-can, sir."

"Puh-khan would be just fine."

I'll give the waitress this: she didn't bend. She didn't break. Somewhere around the fourth or fifth pee-can, she kissed her tip goodbye but she hung in there, pronouncing until her cheeks flushed red. I don't think the service industry was a good fit for her.

"I don't know where you guys are from," she told my father. "But wherever it is, you don't know what you're talking about. The word is pee-can."

The waitress made her last mistake then, mentioning where we were from as if it were the homeland of rubes. My father sought to lock eyes with her, to stare her down. "We're from North Carolina," he said, "and where we're from they have puh-khan trees which is what the puh-khans you've never heard of grow on and which they end up putting in the puh-khan ice cream that you won't serve us. Now what do you think about that?"

What *he* thought about that, I now suspect, was that he was a doctor who'd risen from a small country town in North Carolina through the dint of intellectual application, a man who read Loren Eisley and C. P. Snow for fun and illumination, and here he was having to argue with a roadside waitress over the pronunciation of a word he cherished as a keepsake, as a badge of loyalty to home. He traveled the world hyperalert to any suggestion of slight, to anyone who looked down on the South. And yet the mutinous thought rose into my mind: here was a woman doing to him what he did to us. Telling him how to talk, how to say things. I kept this revelation silent, however. I wanted my father to win. No wonder even iconoclasts fall prey to the seductions of nationalism, of identity politics. You must plug your ears to escape the siren call of home.

"I don't care where you're from," the waitress told my father. "The word is pee-can. For the last time, would you like for me to bring you and your family some pee-can ice cream?"

"No, thank you," my father said. "I don't think we do." He stood up, we all stood up, and we left the restaurant. The pecan ice cream stayed behind, a frozen hostage in the great war of pronunciation. Driving north into the night, my father exclaimed with fury, "That was a *rude* woman."

In Boston, he ratcheted up his courtesy level by a conspicuous degree, mixing his mother-in-law and her sister whiskey sours and holding open doors and bowing his head and saying "Yes, ma'am" and "Yes, sir" to just about everyone who moved.

As an obsessive in the family tradition, I am ruminating about that word—*pecan*—this summer afternoon in New York City. A lawnmower—that gas-fueled engine of nostalgic reverie—roars as the landlord mows his narrow strip of city grass in front of the apartment building. No suburban expanse, no baseball diamond has been more tenderly groomed.

Pecan. How one says things matters. I have no children but I hate the thought that all of my father's foolishness and wisdom and passion, his complicated love, will just disappear into the nullity of time. So here, once again, at the very last they go.

—WILL BLYTHE
NEW YORK CITY
AUGUST 2006

Acknowledgments

THIS BOOK ORIGINATED when its editor observed the writer curled on the floor in what the editor over the years liked to describe with wicked glee as "a fetal position." The writer was engaged in watching a Duke–North Carolina game. The linoleum floor was cold and gritty, paper clips were sticking to his cheeks, and the view of the television was not ideal. Perhaps by now, you understand that by positioning himself on the floor in this fashion, the writer had hoped that he might effect a change in the game's outcome. Although such a result may seem unlikely on the face of it, he believed that one day science would prove what he already knew in his heart: that the passionate interventions of fans like himself could alter the course of games.

"You know what I'm thinking," the editor said that night to the writer. He was wearing the nicest leather shoes.

"That I should write about this," the writer said.

"Exactly," the editor said. "And by the way, you have paper clips stuck to your face."

What began then as the whim of a winter's night soon turned into a collaborative effort the size of a small city. This book would never have been finished but for the wisdom and generosity of a great many people, starting with the editor who asked for it (in more ways than one), my longtime friend, David Hirshey. I am also deeply grateful to Nick Trautwein for his superb editing, unflappable demeanor, and that Miles Davis CD he left at my house. Special thanks in just about every way are also owed to Annie, David, John, and Mellicent Blythe; Mike Carroll; and, last but not least, Erika Mansourian, all of whom listened a great many times over the years to "just one paragraph." I am much indebted to the extraordinary Scott family—especially Melvin, Bridget, and Charles—and to Edgar "Echo" Shelton, Isaiah "Zeke" Johnson, Jide Sidopo, Meredith Smith, and Andrey Bundley.

I am greatly appreciative of the efforts of Sloan Harris and Katherine Cluverius at ICM.

I am also happy to thank the following for their contributions: Sam and Pat Blythe; Sal and Joe Pugh; Doug Marlette; Gretchen Gettes; Louise Hackney; John Charles Blythe; Steve Kirschner and Matt Bowers at the UNC Sports Information Department; Jon Jackson, Matt Plizga, and Melanie McCullough at the Duke Sports Information Department; Terry McDonell, Rob Fleder, and Dick Friedman at *Sports Illustrated*; Jim Young of the *Greensboro News & Record*; Mark Warren at *Esquire*; Tom Colligan; The North Carolina Collection at the Wilson Library at the University of North Carolina; Tim Pyatt, Tom Harkins, Jill Katte, and Nancy Perry at the Duke University Libraries Archives; Ben Sherman and Jim Hawkins at *Inside Carolina*; Mike Hemmerich and colleagues at the *Duke Basketball Report*; Wayne Gooch; Pasha Majdi; Bucky Waters; Matthew Laurence; Sean Dockery; Luol Deng; JJ Redick; Trajan Langdon; Charles Kunkle; Ted Moore; Philip McLamb; David Noel; Jackie Manuel; Marvin Williams; Bill Guthridge; Phil Ford; Joe and Matt Holladay; Eric Hoots; JamesOn Curry; Coach John Moon; Felton and Joan Brown; Tim and Laurinda Krotish; Dawn Allen; Christopher Fordham; William Friday; Dick Baddour; Armin Dastur; Jennifer Holbrook; Laura Altizer; Cathy Mann; Carolyn Webb; Mary Lynn Warren; Eugene Corey; Elliot Warnock; Barry Lawing; Kathleen Mays; Sherrell McMillan; Gil Schwarz and Lesley Ann Wade at CBS; Meg McVey; Rocky Rivero; Leslie Kroeger; Sally Sather; Bill Brill; Robert Seymour; Eryn Wade; Adam Cohen and

Distill; Bill Tonelli; Mark Jacobson; Rob Trucks; Bill Farley; Judy Hirsch; Anton Turkovic; the Cadillac Man; Mark Clein; Daphne Athas; Rodney Peele; and Mark Bryant.

Among the books that I found vital were the following: Fred Hobson's magnificent study, *Tell About the South*; William Snider's *Light on the Hill*; Lindley Butler's and Alan Watson's *The North Carolina Experience*; James Smylie's *A Brief History of the Presbyterians*; Philip Fisher's *The Vehement Passions*; William Hazlitt's *On the Pleasure of Hating*; Joe Menzer's *Four Corners*; Robert Durden's *The Dukes of Durham 1865–1929* and *Bold Entrepreneur: A Life of James B. Duke*; Bill Brill's *A Season Is a Lifetime*; David Halberstam's *Playing for Keeps: Michael Jordan and the World He Made*; Barry Lawing's *Demon Deacon Hoops*; Dean Smith's *A Coach's Life*; Mike Krzyzewski's *Leading with the Heart*; Barry Jacobs's *Coach K's Little Blue Book*; William King's *If Gargoyles Could Talk: Sketches of Duke University*; and James Applewhite's wonderful poetry, starting with *Statues of the Grass* and including his recent *Selected Poems*.

No book about hatred could have been more a labor of love, and that is due to the kindness of those listed. Again, my thanks.

CPSIA information can be obtained at www.ICGtesting.com
Printed in the USA
LVOW08s1422230915

455408LV00031B/269/P